Thomas Edmondson and his Ticl

To Len Peach with many thanks from one "squirrel" to another!
Michael Farr 30.OCT.95

by Michael Farr

*This cartoon appeared in **Print,** the journal of the National Graphical Association in a series titled **The History of Print.** Thomas Edmondson would, no doubt, have been proud to know that in 1990 the systems he pioneered and the machine he invented were considered to merit a place in printing history. (Reproduced courtesy: NGA)*

British Library Cataloguing in Publication Data
Farr, Michael, *1938-*
Thomas Edmondson and his tickets.
I. Title
358.262
ISBN 0-905033-13-2

Published by the author at
40 Lillywhite Crescent, Andover, Hampshire. SP10 5NA

First printing 1991

To Pauline,
25 years a tickets widow.

Grateful thanks for assistance with production to Tracey Daykin, Janine Golding, Kevin Holmes, Amanda and Peter Taylor and Richard Unthank.
Typeset by the author and Anton Graphics Ltd, Andover, reproduction by Oxford Litho Plates Ltd, printed by John Dollin Marketing Services Ltd, Whitchurch, Hants.

Foreword

MYSTERY always surrounded life behind the ticket counter, for little could be gleaned from peering through the tiny keyhole window into the dark recesses of the traditional railway booking office.

Perhaps it was that mystery that caused my lifelong interest in tickets and fare collection. On my daily journeys to school I (and other regular passengers) would be invited to warm ourselves around the booking office fire whilst waiting for the 8.20 a.m. from Sea Mills to Clifton Down, Bristol. I suspect the porter in charge (Sea Mills did not rate a full station master) was only too happy for this keen ten year old to date and sell the tickets and even to help with the task of booking-up. How sad that today's need for security in offices handling money prevents such trust - and such visits. For in those days who would have contemplated betraying that trust and helping themselves to tickets or the till?

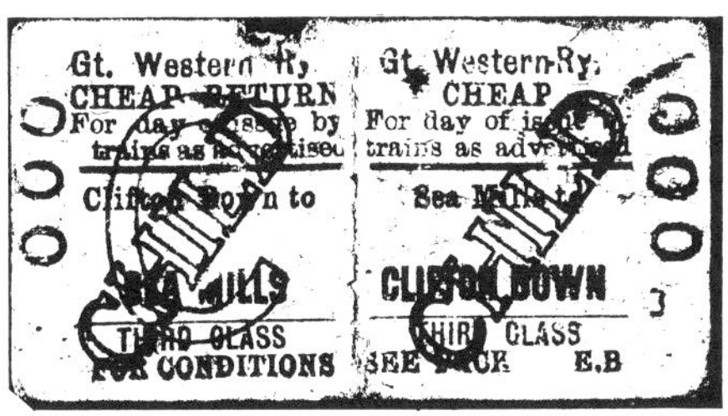

▲ *01 JUN 48* *25 OCT 50* ▼

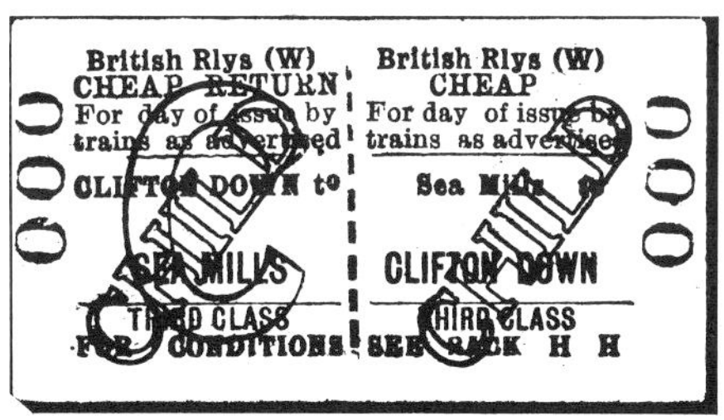

It was here, too, that I purchased the first ticket to start my collection more than forty years ago. On 1st June 1948 post-war travel restrictions eased to allow the general reintroduction of cheap day return fares, and the cost of my journey to school reduced to 3d return. That morning Bill White, the porter, sold me the first ticket of the series, numbered 000, so I invested a second 3d to ride and kept 000 to begin my collection. I have it still, as well as the 10,000th and 20,000th tickets issued to me for the same journey over the ensuing seven years. When the series reached 9999, they always reverted to number 000.

After that a call at the booking office was essential every time we passed a station, so I could buy a platform ticket (1d) or the cheapest child single to the next station. I spent many happy holidays begging to be allowed into offices in Cornwall, Devon and Somerset where I was able to buy a Holiday Runabout Ticket for a few shillings to visit places no longer served by rail - like Helston, Bideford, Hemyock and Polsham Halt.

A chance remark by one of the railwaymen who bent the rules to allow me to pass behind the PRIVATE sign about having seen at Lancaster station a plaque to the memory of Thomas the ticket man was all the fare I required for my journey of discovery.

It has taken me all over the UK, to correspondence with rail officials across the globe and, the highlight, to the SBB ticket works at Bern, spotlessly clean as might be expected. The Swiss printers could speak little English; I could speak no German, but the ice was broken when one of their number presented me with a line of type set up to read - **EDMONDSON** - and a rapport was immediately forged!

I received the rare privilege of a visit to British Rail's ticket printing shop in 1970 and was fortunate to return several times over the ensuing twenty years. The staff were always ready to explain processes and answer innumerable questions, even when threatened with the department's closure. I must mention especially Cliff Dobson, Geoff Bavington, Mick Norman and Ron Johnson, as well as Joe Turner, who masterminded the amalgamation of ticket printing after nationalisation, and his successor, George Bryant, who was a mine of information about GWR printing at Paddington.

During my researches I have learnt much from railwaymen of all grades, printers and their staff and particularly fellow members of the Transport Ticket Society. I must record special thanks to David Geldard for permission to use information from his book *The First Fifty Years*. Brian Pask, John Tilly and John Shelbourn have kindly agreed to read my manuscript and their comments, gratefully received, have been incorporated into this edition.

Rev. Wilbert Awdry not only explained the ticket practice on his fictitious Skarloey Railway, but also kindly put at my disposal his records of the Birmingham and Gloucester Railway and two of its employees - Thomas' brother and son.

Many operators and firms supplying ticket printing machinery and tickets themselves have patiently answered my questions and searched dusty archives. Red Funnel Ferries discovered Waterlow's guard books of early tickets and Glasgow Numerical Printing Ltd even presented me with their Waterlow ticket printing machine, which is now restored and on display at Amberley Chalk Pits Museum, West Sussex.

My wife and family have been exceedingly tolerant of 25 years of ticket collecting. They have learnt to disown me when receiving blank looks at the booking office after asking to buy specimens "just for my collection" or even searching used ticket boxes and bins for clean discarded examples. I hope they will now consider it all worthwhile.

We are always learning, and I shall be grateful for any further information, views (or unwanted tickets!) that readers may have. Correspondence following the first editions forged links with relatives of both our hero, Thomas Edmondson, and his business colleague John Blaylock.

A recent contact is Revd Alan Cliff, who is researching in detail the development of railway ticket dating presses. This is, I feel, a specialised subject which deserves more depth than I can give in this slim volume, and I look forward to the publication of the fruits of Alan's research.

Thomas Edmondson was one of the most unlikely entrepreneurs of the Victorian era. His invention has been compared with that of Sir Rowland Hill — the adhesive postage stamp — whose early development coincided with the first card tickets. There the similarity ends, for with millions of different varieties of ticket printed during the past 150 years or so, it is most unlikely that a future Stanley Gibbons will attempt a definitive catalogue of railway tickets, though the Transport Ticket Society does publish an excellent series of specialised papers.

. . . and that unknown factor of what you may discover next is another appeal of collecting tickets.

Andover, Hants. MICHAEL FARR

Thomas Edmondson Transport Ticket Pioneer 1792-1851

Two separate plaques in Lancaster, one at the site of his birthplace and one at the railway station, commemorate a one-time journeyman cabinet maker who was to pioneer a system of fare collection in the early days of railway development. So simple and efficient were Thomas Edmondson's methods that their principles remain in daily use with transport undertakings all over the world. Indeed, for more than 150 years British railway stations in many parts of the country have held a stock of cardboard tickets, size 2¼ x 1$\frac{3}{16}$in (57x30mm), the same dimensions as introduced by Edmondson in the first years of Queen Victoria's reign.

He was born in a house on the corner of Stonewell on 30 June 1792, one of twelve children of John and Jane Edmondson, members of the Society of Friends. Only five of their children reached maturity, three boys and two girls. Their parents decided that the boys should be encouraged to follow the career for which they seemed best suited, so Thomas' two brothers (Joseph and George), who showed academic ability, became schoolteachers. Joseph gave up teaching in middle life, but George built up a high reputation as an educationalist.

> *"Oakrigg," Lancaster.*
> *1st August, 1930.*
>
> The Memorial Tablet on the Birthplace (over the shop of Mattie Clucas') in Moor Lane, Lancaster, of
> **THOMAS EDMONDSON,**
> the Inventor of the Railway Ticket Dating Press and Printing Machinery, will be Unveiled by his Grandson (Mr. Thomas J. Edmondson, of Manchester), on Thursday, 7th August, 1930, at 12-30 p.m. The pleasure of your attendance is requested.
>
> T. Cann Hughes.

He was invited in 1817 with a friend Daniel Wheeler by Tsar Alexander I to oversee agricultural projects in Russia, returning to England in 1830. After teaching at Blackburn and Preston, he moved south to Queenwood College, Hampshire. Here he took over Robert Owen's former Harmony Hall and developed it as an agricultural, scientific and technical school. He is reputed to have set mathematical work for his students based on Bradshaw's railway timetables!

Thomas on the other hand, had more of a practical nature and was always working with his hands. He connected the baby's cot to the butter churn, so butter could be made as the cradle was rocked! It was appropriate, therefore, that he should serve a cabinet maker's apprenticeship, first with a neighbouring woodworker and later with the famous Lancaster furniture-making firm of Robert Gillow and Company. He completed his apprenticeship with them, becoming a journeyman, and his son recorded in a biography written in 1878 that while with Gillows Thomas Edmondson had suggested various improvements to tools, which his fellow craftsmen accepted as a great advance.

To Cumbria

Being ambitious, he moved to Carlisle to become a partner with friends in a cabinet making business, but this unfortunately lost money and he was declared bankrupt. Due to his deep religious convictions, Thomas' debts concerned him greatly and he and his family had to leave the Society of Friends for many years. He vowed one day he would repay his creditors, but little did he think then that a small piece of printed cardboard would enable him to do so - and in due course to be welcomed back into the Friends shortly before his death.

In an effort to improve his financial matters, he entered the tea and grocery business, work for which he felt he was not suited. Thomas Edmondson then applied, in 1836 at the age of 44, to join the Newcastle and Carlisle Railway; his application was successful, the N&CR directors minuting their opinion that "Mr. Edmondson would be a credit to them", although they later did little to show their appreciation of his abilities.

According to Alfred Morris, curator of Naworth Castle, Brampton, Edmondson was first appointed station master at Scotby, but he must have quickly moved to Milton, about fourteen miles from Carlisle. In his biographies published in catalogues and press articles his son, John B. Edmondson, mentions only his service at Milton station, which was subsequently renamed Brampton Junction and is now known simply as Brampton.

By the end of the 1830s some 50 railway companies were operating passenger services. A pioneer line, the Leicester and Swannington, introduced in about 1832 brass octagonal checks punched with various information, including a serial number. These were used to control passenger priority on trains as well as for accountancy. The checks, collected at the end of the journey, were returned to the booking office for reissue, as were similar (but unnumbered) copper checks used on the London and Greenwich Railway.

Everyday ticket practice on other lines generally relied on paper tickets, torn from books, various details being entered by hand at the time of issue. The S&DR ticket below, printed in black ink on white paper, measures 105 x 70mm. The lack of pre-printed serial numbers opened the door to fraud and, when (as on the N&CR) the station master collected fares and passed them to the guard of the train to pay in at the terminus, there was no satisfactory way to check that all monies duly reached the company's coffers. This loophole concerned Thomas Edmondson with his high Quaker principles.

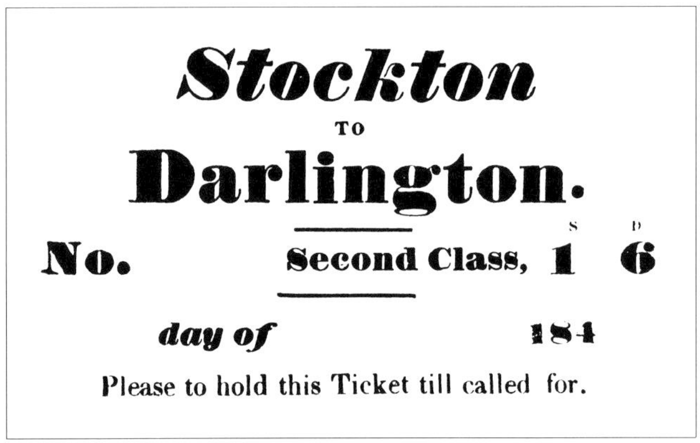
Stockton
TO
Darlington.
No. Second Class, 1 S 6 D
day of 184
Please to hold this Ticket till called for.

After considering the problem, Edmondson evolved a method of preparing in advance receipts in the form of card tickets, which each passenger could be given in exchange for his fare, with three main objectives:

- *to ensure the passenger paid the fare.*
- *to establish his entitlement to travel between two points.*
- *to ensure that all money due was paid to the operating company.*

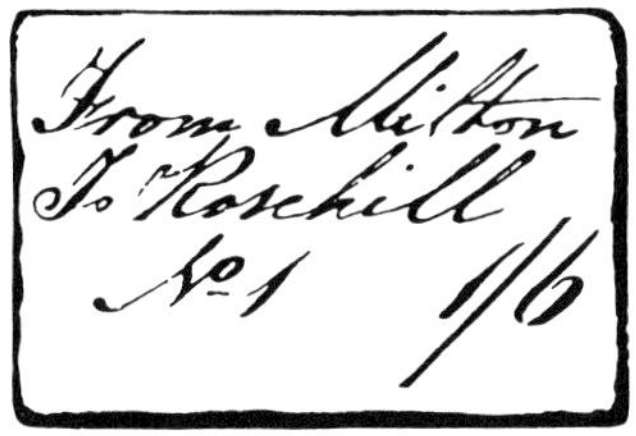

Edmondson took a strip of card and wrote by hand several times **MILTON TO CARLISLE, FIRST CLASS, 1/-.** He then numbered each ticket individually and after cutting the strip into separate tickets he was ready to handle the bookings.

By keeping a note of the serial number of each set of tickets at the start of every day, subtracting that from the serial number at the end of the day and multiplying the resulting total by the fare, he could work out precisely how much should be in the cash till.

An important point was that the first ticket of each series was numbered "0" and not "1", so that the number of the next ticket to be issued represented the total of that type of ticket so far issued. Had he begun numbering at "1", it would have been necessary to subtract one.

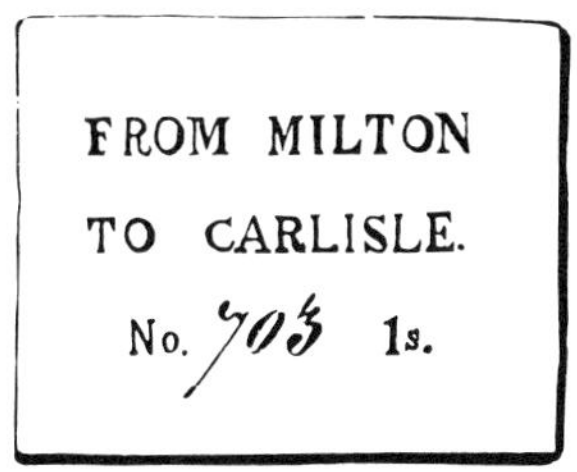

To write each ticket by hand had proved a laborious task, so Thomas Edmondson made a simple printing frame on the principle of a modern rubber stamp and ink pad. His son later recorded that he set up type, although examination of extant tickets suggests from the variation in lettering that he carved woodblocks by hand, giving the station names, class and fare. This is not surprising in view of his expertise with wood. He then placed a strip of card in the frame, inked and positioned the block, and a sharp blow with a mallet made the impression.

Whilst his colleagues at other Newcastle and Carlisle Railway stations were keen to adopt the edmondson system, the directors held back, and promises to promote Thomas to a more responsible position at Newcastle never materialised.

Overtures from Manchester

When, therefore, Captain Laws of the Manchester and Leeds Railway visited Milton to see the system for himself and offered Thomas a post at Oldham Road, Manchester, at double his N&C salary, he was glad to accept and moved south in May 1839.

Whilst with the M&LR (opened on 3 July 1839) he was able to introduce many refinements and left that railway in 1841, with a complimentary minute of the board. He then established in business his son John Beeby (whose date of birth is recorded as 22 June 1831) and his brother Joseph (who was to manage the firm until J. B. came of age) to produce tickets and associated machinery. The company, which traded in Manchester with offices in London and Dublin, went under the title of John B. Edmondson Limited.

The Birmingham and Gloucester Railway adopted the Edmondson system in 1840 and his brother and (at a remarkably early age) his son were employed by the B&GR to initiate the operation. By August 1841, fourteen English companies and the first overseas railway, the Paris and Rouen, were using the system. By

March 1843 the list included a further thirteen English companies and from that time it became the standard system to adopt.

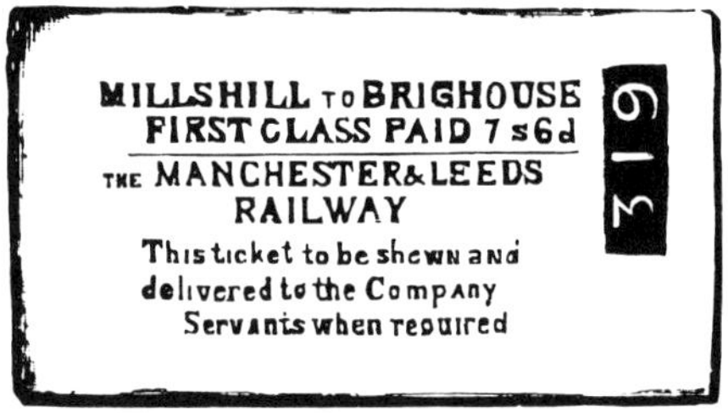

The Railway Clearing House was set up to administer (among many other matters) the crediting of each railway with their dues from through tickets and consignments over two or more separate companies' lines, and it became virtually a condition of membership that the edmondson system be used.

When the B&GR began an economy drive in 1842 John left their employ, with Joseph following at the end of 1844, to work with Thomas. For the privilege of using the edmondson system, a railway was charged a royalty of 10/- (50p) for every mile of track. Thomas requested also a free pass to travel over the system, although the London & Birmingham (who introduced his tickets in August 1841) were apparently a little reluctant to agree to this. A letter displayed at the GWR Museum at Swindon from Edmondson to the L&BR offers to review this clause in the agreement to hasten its signing.

Dating the Tickets

Tickets are usually valid for a prescribed limited period varying from one day to three months or even more. To prevent fraudulent use, therefore, it is necessary to print the date of issue on each ticket. Thomas Edmondson is reputed to have conceived an idea for the nipping action of his "pendulous" dating machine from his folding pocket comb. The date is set in printers' type and when a ticket is inserted in the jaws of the machine, gentle pressure enables the type to print the date through an inked ribbon.

One of Edmondson's earliest wooden dating presses was displayed at the Exhibition of the British Association in Newcastle in September 1838. Examples of these machines are on view at Lancaster City Council Museum, the Science Museum, London and the National Railway Museum, York and others are in private collections, a testimony to the quality of workmanship. The *Illustrated London News* engraving below shows a ridiculously under-scaled hand!

The dating machine was subsequently developed into the iron press once to be found on almost every British booking office counter, and in many overseas countries. The prototype was made by Ralph Cairns, a Brampton clock and watchmaker, but for the bulk production Edmondson turned to John Blaylock, who employed Cairns' son.

Blaylock, originally from Dublin, operated a watch and clock business in Carlisle, and had constructed the unusual five-faced clock which once hung at Carlisle Citadel station. He also ran a foundry producing domestic ironware, where the castings for the dating presses were made. Assembly provided employment for former clockmakers whose eyesight had deteriorated, preventing them from undertaking intricate work - thus fulfilling a social need. In 1859 William Pratchitt joined Blaylock and in time the company became known as Pratchitt Brothers Ltd.

The type used in the metal dating presses is the old Great Primer size, superseded in the printing industry by the point system of sizing measurement at the end of the last century. Although the type body size remained the same (to fit the press) different railways used different type faces, amongst them Roman No 6 and Pica Antique No 2 by V. J. Figgins (later R. H. Stevens' foundry) and Great Primer No 10 from Stephenson Blake. The type sometimes carries the founders' name, cast on the side.

Type was usually cast in founders' metal, although brass type (for longer wear) was sometimes provided and London Transport latterly issued steel type, engraved with the days of the month (-1 to 31) and the month abbreviation, e.g. JAN, FEB.

Some later models impressed the date into the ticket card, making it almost impossible to alter the date after the ticket had been issued. One design printed and impressed the date, while another had a facility for keeping a variety of different dates set up in slides, so that tickets could be issued in advance at agencies such as Thomas Cook Ltd.

Edmondson's dating presses usually carry a brass plate quoting his name, and sometimes Blaylock's as well. Several other firms marketed ribbon-inked presses including Adams & Whiteman, Waterlow (some models attributed to the patented design of W. T. Whiteman) and Williamson. Many Eastern and North Eastern stations used date presses carrying a plate "The Eldon Engineering Co., Elswick Court, Newcastle-on-Tyne" and London Transport latterly obtained their presses from Malcolm Clifford Ltd of Ilford, after Waterlow ceased to handle them.

JOHN B. EDMONDSON, MANCHESTER,

RAILWAY TICKET DATING PRESSES

(THOS. EDMONDSON, INVENTOR).

ORIGINAL RIBBON PRESS.

(ORIGINAL PATTERN PRESS).

PRICE COMPLETE ON APPLICATION.

TYPE BOX WITH TYPE SUPPLIED WITH EACH PRESS.

The Ribbon Dating Press, although having been used for a great many years, has been very little altered, and may be said to be the best press for ordinary purposes. It is very simple in construction and seldom wants repairing.

IMPROVED RIBBON PRESS.

(FOR RAPID CHANGING OF DATES.)

PRICE COMPLETE ON APPLICATION.

TYPE BOX WITH TYPE SUPPLIED WITH EACH PRESS.

As illustrated above.

The advantage the Improved Ribbon Press has over the original one is that it enables the operator to have several dates set up in holders as above, any of which can be used, and changed again in a very short space of time. This press is specially convenient at stations where large numbers of excursion tickets are issued, or where tickets are dated in advance of current date.

Repairs to Dating Presses receive prompt attention, and the press is returned as good as new. Dating Press Ribbon supplied either inked or uninked of any length.

LONDON, GLASGOW, DUBLIN.

Other designs of press featured integral printing wheels which could be set to indicate day, month and year, and were inked from a pad instead of a ribbon. The LNWR used pinhole dating at the turn of the century, although internal correspondence which survives suggests the machines were far from reliable. The figures and letters were punched out completely, a system used to validate coupon-type international tickets at travel agencies for many years.

Waterlow continued to offer an impression dating machine, which was popular in tropical climates (where an inked ribbon would presumably dry out and the security of an impressed date was considered necessary) and their successors, ERA Limited, received an order from the middle east for such daters in the late 1970s.

Several continental railways still (in 1991) impress the date on edmondson cards using desk-top machines, many in Switzerland engraved with the name J. J. Güller. Güller Sohne AG, a company founded in 1845, is still active in Hüttikon; its 1905 catalogue illustrated several impressing and perforating dating machines, as well as ticket clippers, station stamps and serial numbering machines. The model illustrated *(left)* appears very similar to the final version of Waterlow impression dater, as made by ERA Limited.

▲ *Güller Datum-presse from 1905 catalogue.*

▼ *Frontispiece to J. B. Edmondson catalogue.*

ALL APPARATUS, AS ILLUSTRATED, MANUFACTURED

JOHN B. EDMONDSON'S

Railway Ticket and Apparatus

CATALOGUE,

INCLUDING THE LATEST IMPROVEMENTS IN

Railway Ticket Printing, Backing, Counting, Dating, Issuing, and Destroying Machines,

INVENTED BY THE LATE

MR. THOMAS EDMONDSON,

ORIGINATOR AND INVENTOR OF THE RAILWAY TICKET SYSTEM.

Knowsley Street, Cheetham, Manchester.

LONDON:	GLASGOW:	DUBLIN:
Holford Square, & Pentonville Rd., N.	117, Centre Street, S.S.	Alexandra Buildings, 12 & 13, Eustace St.

AT MY ESTABLISHMENT IN MANCHESTER.

Printing by the Millions

Edmondson and Blaylock now turned their thoughts to mechanising the printing of tickets, as the hand operated "frame and mallet" system was obviously inadequate to meet the needs of the expanding railway network.

With Edmondson's ideas and Blaylock's metal-working experience they produced a prototype machine for printing and numbering tickets consecutively, inked by a wide ribbon impregnated with ink. This is to be seen at Lancaster City Council Museum, to whom it was presented by John B. Edmondson Ltd in 1941, the contact between the City and the Edmondson company having been forged in 1930 when the plaque on his birthplace was unveiled.

In preparing his paper *Railway Ticket Printing and Numbering Machine*, Robert J. Excell of the Science Museum examined the Lancaster machine (although he was unable to operate it) and concluded that it had been designed to print tickets in Edmondson's early size, 1⅛ x 1½in, smaller than the standard 2¼ x 1³⁄₁₆in later adopted.

An outstanding feature of the hand operated machines is a pair of large brass wheels, turned with a pawl and ratchet, on which are engraved the serial number figures (reversed-out white on a black square) 00 to 99. With every stroke of the handle one wheel progressed to the next number; when 99 was reached a carry mechanism moved on the hundreds/thousands wheel.

A production model of this machine was featured in the *Illustrated London News* dated 22 February 1845, from which the illustration *(right)* is taken. An example was presented by the family firm to the Patent Office Museum in February 1869. This is now exhibited at the Science Museum, London where it has been

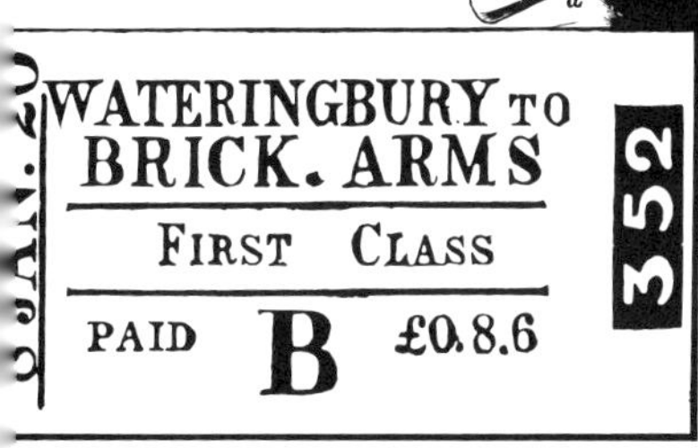

used occasionally to print tickets for Museum events, such as for the demonstration runs of the replica broad-gauge locomotive *Iron Duke.* The Museum retains the original correspondence with John B. Edmondson in which it is recorded that the printing machine was installed by Edmondson's London agent, Mr Beckett of 109 Pentonville Road.

This later design of machine was covered by Edmondson's patent No. 8538, enrolled on 9 December 1840. His next patent for improving the machine - printing with ink in place of ribbon, automatically stopping the machine on completion of the required quantity and driving it by power - was granted (No. 12137) on 27 October 1848. The machines were now approaching their final development offering printing speeds of up to 12,000 per hour, necessary as the number of passengers travelling escalated, with further improvements through his patent No. 13007 of 14 September 1850.

John B. Edmondson, Thomas's son, who took over the Manchester company, clearly had a flair for publicity and illustrated opposite is a typical title page from his catalogues which always included a biography of his father. He expanded this biography to describe in an 1878 issue of *The English Mechanic and World of Science* first the early development of the system and then the operation of the later powered ticket printing machines, as follows:

"The printing machine . . . is put in motion when a catch, set to nearly the thickness of a ticket, and working horizontally, draws the lowest card (in the feeding tube) in the direction of the type and numbering wheels . . . If it is to be a return ticket it receives a number; if not . . . the first set of wheels has been raised and the ticket waits to be pushed forward to the type by the introduction of a second card from the feeding tube.

"A third card pushes number one under the second set of wheels, where it receives its appropriate number, and by the push of the fourth it falls, printed and numbered into a receiving tube at the front of the machine . . .

JOHN B. EDMONDSON, MANCHESTER,

RAILWAY TICKET PRINTING MACHINE

(THOS. EDMONDSON, INVENTOR).

PRICE ON APPLICATION.

My Railway Ticket Frinting Machines are made of the best material and workmanship, on my premises in Manchester. They are largely used for printing tickets for British railway companies, and the supply of tickets printed by them has given unrivalled satisfaction. The above Machine is capable of printing and consecutively numbering the large quantity of 12,000 tickets per hour, and can be driven by either power or hand. The working of it is an easy matter, and can be done by a youth of ordinary intelligence. It only requires feeding with blank, and relieving of printed tickets, the machine accomplishing all other operations. My machine is supplied with specially designed block numbers, or raised numbers, the former being generally chosen as they are much more difficult to forge.

LONDON, GLASGOW. DUBLIN.

"The mechanism, while capable of being driven at a great speed, is regulated to that of about 200 tickets a minute, this being found a rate at which the attendant can most readily superintend the supply of blank cardboard to his feeding tube and give . . . attention to other movements of the machine".

One Edmondson patent included the facility to print a company's monogram or other device on the back of a ticket. Because some railway staff and many passengers were unable to read the station names on a ticket, on the early Manchester & Leeds Railway tickets distinctive patterns were printed on the back of the card to indicate the different stations. Several railways did print their monogram *(Edinburgh & Glasgow above)* and some Indian tickets still show the company initials on the back for added security.

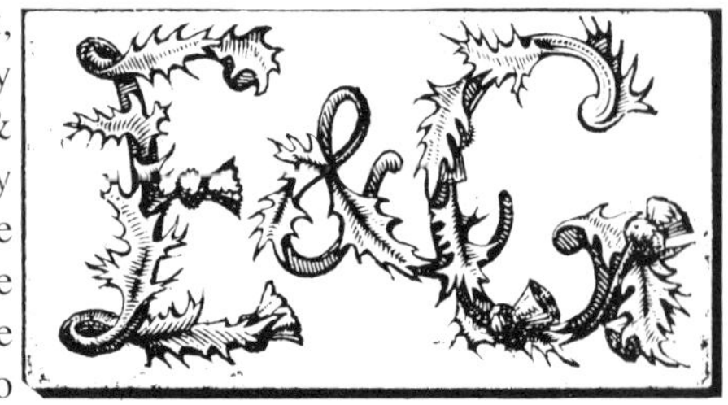

Inspecting and Clipping the Tickets

There is much dispute amongst those who collect tickets as a hobby whether they are more desirable when clipped, or not. According to James Mackay in *Railway Antiques*, "Americans claim that the ticket-punch was invented by conductors on the Erie Railway, but the British give credit to William Carson of Creetown in Scotland, who was ever after known as *Tickety* Carson". It appears that Carson and Edmondson collaborated on the introduction and distribution of clippers, and the former John B. Edmondson catalogue listed and illustrated 33 different shapes of clip. The intention was that from the different clips ticket inspectors were able to discover what portion of a ticket had been used, important when passengers took advantage of the option to break their journey.

British Railways' manuals illustrated dozens of different clips and marks, with a punched shape often being combined with an embossed letter/figure code. A German firm offers no less than 1,000 different clip marks in its catalogue.

The Great Western Railway had for many years provided stations with clippers which removed a small hole and underneath this impressed a code so the precise station of clipping could be checked. Even some cut-out clips could be related directly to a station, with Brighton having a B shape, Portsmouth and Preston a P, Reading an R, London Victoria a V and Waterloo a W. Shipping services had a ship outline, which continued in use with Sealink after the railways hived off their ferries.

In the mid 1980s British Rail began on many routes to phase out the traditional clipper in favour of Dormy self-inking cancellors which printed seven variable digits to indicate the date and route, as well as a fixed reference code. One reason for the change was to avoid problems which would have been caused later if clipping with a traditional punch had damaged the coded magnetic strip now used to operate entry and exit barriers at London Underground stations.

For many years most large stations were "closed" with ticket barriers manned by staff whose duty it was to examine or collect tickets. Finally, after the ticket had been surrendered a cancelling clip was made (usually a V shape) to prevent its further use.

A development of the closed station was the "Crewe" system, a reminder of the days when each railway company had its own offices at a major jointly-operated junction station. Here the various railways' ticket offices were situated on the platform, so it was necessary to enter the station past the inspector's booth to purchase a ticket. Intending passengers were given a pass, usually a standard ticket but at a few stations a re-usable metal token, which they surrendered on payment of their fare. Crewe was the most famous station to operate in this way, although there were several others.

Another variation was at Taunton, which formerly had separate offices for up and down bookings on opposite sides of the station. The enthusiast soon learnt that if you asked for a ticket to an up destination at the down office you would receive a special free platform ticket to pass through the subway to the other office; thus train spotting could be free, so long as you remembered to leave the station by the opposite exit! Many other stations held a supply of free platform

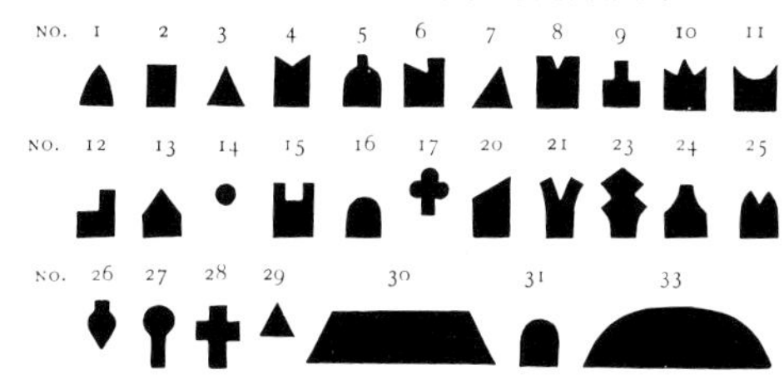
JOHN B. EDMONDSON, MANCHESTER,

RAILWAY TICKET NIPPERS.

SPECIMEN OF VARIOUS CUTS MADE BY NIPPERS.

The Ticket Nipper is chiefly used to show that the railway ticket has passed examination at a certain station, or to cancel the ticket altogether.

I make a special Railway Ticket Nipper, with moveable type, to indent the current date into the ticket, each being supplied with a type to cut a private mark, used by inspectors.

tickets, usually at the barrier, to issue to visitors wishing to use the left luggage facilities or refreshment rooms which were usually on the platform. The author was granted a free annual pass at Temple Meads Station, Bristol, to enable him to use the refreshment rooms at lunchtime.

A clipper with an entirely different purpose was known as the half-snip. Where sufficient numbers of children travelled between two stations, a specially printed ticket was held for issue at the lower (usually half) fare. However child tickets could also be issued by using the half-snip to remove a large semi-circle, triangle, rectangle or trapezium from the base of an adult ticket. On the piece removed was printed in code enough information to assess the fare and credit the booking clerk with the value of the snip.

Many pre-1922 companies used this system, as did the LNER and LMS until 1947. Indeed, pre-printed stocks of child tickets on the LMS were supplied from the printers with the snip already removed. Other transport operators which used the half-snip system were the Isle of Man Railway, Isle of Man Steam Packet Company and the General Steam Navigation Company. Some early Talyllyn Railway tickets carried the small repeat of the destination centrally at the bottom of the

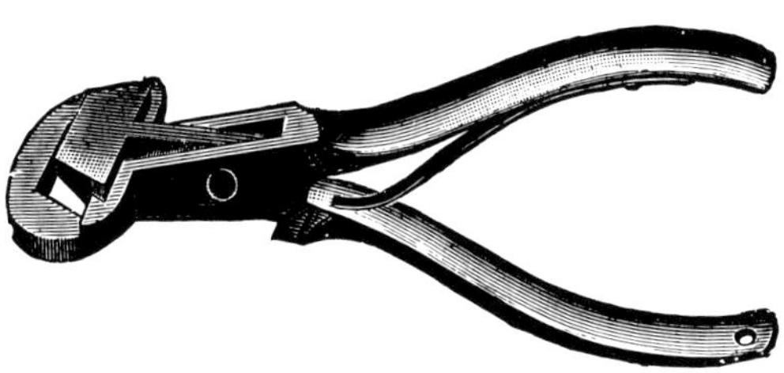

For children's or half fare tickets.
To cut half round or oblong.

Some halved tickets for children.

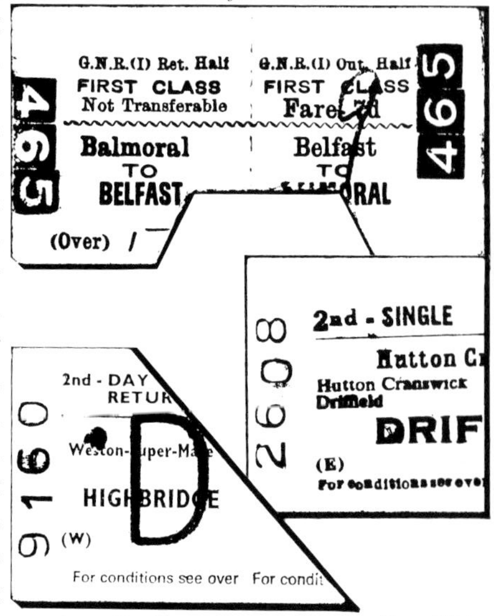

ticket, but there is no evidence that half-snip clippers were used there.

An alternative system depended on the fact that the child fare was usually exactly half that for an adult. Thus, for example, at a GWR or SR station when a child tendered his fare an adult ticket would be taken from the rack, dated and cut in half; singles were cut vertically, returns diagonally. This system was particularly valuable at Temple Meads, Bristol which held stocks for a vast range of destinations; by concentrating on adult tickets much space could be saved.

At the end of the month all odd half tickets remaining were cancelled and the station credited with their value. This system became the standard for all four companies (and subsequently BR) from 1947. Standing instructions forbad the cutting in half of "blank card" tickets, used for little-used journeys and completed manually by the clerk . With these children received the whole ticket, rubber stamped **CHILD**.

Another use for the half-snip clippers was to remove a portion at the top of a forces leave or privilege ticket to indicate that it could be used only by a female passenger.

Dispensing the Tickets

Storage was a problem with the small individual bundles of card tickets, and Edmondson used his joinery expertise to construct a rack of "fuses" for dispensing them one at a time, as well as drawers and cupboards for storing them in bulk. With the idea of having the next-numbered ticket always in view he initially arranged to take it from the top of the rack. He made tubes with loose bottoms, forced towards the top of the tube with tapes, pulleys and weights, on a similar principle to a sash window.

This proved cumbersome, so he rearranged the tubes to allow the tickets to drop by their own weight, being removed in numerical sequence from the bottom of the tube. A strip of slate was often fixed in front or on top of the tubes so the initial serial number could be noted before issue and also a strip or two of paper to carry destination details and prices of individual tickets.

On appropriate railways a slot or wooden strip was also positioned to hold the unissued halves of child tickets. When the LMS abandoned the practice of removing half-snips, stations were issued with special brass clips to attach the half tickets to the front bars.

The workmanship of ticket cases was invariably of the highest order, often with either roller shutters or with doors containing further ticket tubes designed to lock the whole ticket stock securely. Racks of similar design but with wider divisions were provided for holding and dispensing the larger sized season tickets. Thousands of racks of this design remained in use at British Rail stations for almost 150 years, and hundreds can still be found dispensing edmondson style tickets on many independent and preserved lines.

JOHN B. EDMONDSON, MANCHESTER, 13

TICKET ISSUING CASES

(THOS. EDMONDSON, INVENTOR).

CASE No. 1 ON THE LIST. (SEE PAGE 14).

DIMENSIONS.—CLOSED.
Width ....20 inches.
Depth .... 7 ,,
Height ....25 ,,

DIMENSIONS.—OPEN.
Width ....39 inches.
Depth .... 7 ,,
Height ....25 ,,

PORTABLE TICKET CASE
(To hold 4000 Tickets).
PRICE ON APPLICATION.

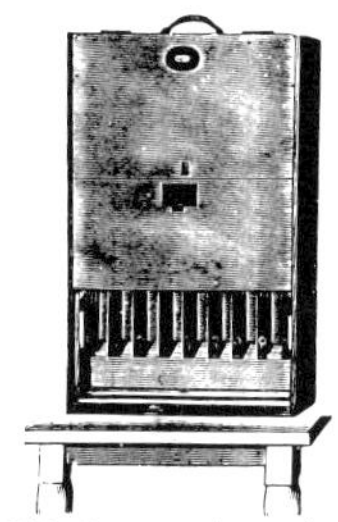

This Case contains 8 tubes, each holding 500 tickets.
SUITABLE FOR PRIVATE USE.

SMALL TICKET CASE.
(To hold 900 Tickets.)
PRICE ON APPLICATION.

This Case contains 18 tubes, each holding 50 tickets.
SUITABLE FOR PRIVATE USE.

TIN TICKET ISSUING CASES. FOR CARDBOARD OR PAPER TICKETS. ALSO SUPPLIED.

LONDON, GLASGOW, DUBLIN.

Alternative Sources of Supply

When Thomas Edmondson died at his home, Meltonville, Crumpsall, Manchester on 22 June 1851 he had the satisfaction of knowing his ticket system was in universal use. The firm's catalogue listed machinery to print the fronts and backs of tickets, to number and perforate them and to check the numbering was in sequence. It offered cupboards and drawers to store bundles of tickets, racks of all sizes to hold them ready for issue, dating machines, nippers to punch them and, finally, a shredding machine to ensure they could not be re-used.

The firm of John B. Edmondson Ltd continued to produce tickets until its director, Mr W. H. Jones wrote to their customers after the second world war to tell them that "... we are closing down our ticket printing department ... As you are no doubt aware, this Company were the originators of the Railway Ticket System, over a century ago."

Among other companies providing traditional edmondson style tickets and equipment were Alfred Williamson Ltd of Ashton-under-Lyne, Glasgow Numerical Printing Company Ltd (GNP), M. Harland & Son Ltd. of Hull, McCorquodale & Company of "Newton on the Liverpool & Manchester Railway" (to quote from their letterheading), George and Alex Falkner of Manchester and Waterlow & Sons of London Wall. Most were old-established firms, Williamsons claiming to have been founded in 1835.

Glasgow Numerical Printing (known as GNP) was formed by David Carlaw in 1878. As an instrument engineer he worked with Lord Kelvin but became interested in printing machines, and in tram ticket printing in particular. When he could find no printer willing to instal the machine he had invented to produce and serially number punch type tickets, he set up in business with three of his ten children.

David Carlaw & Sons' building in Finnieston Street, Glasgow, as it looked at the beginning of this century, from an old letterheading. (Courtesy: GNP Ltd.)

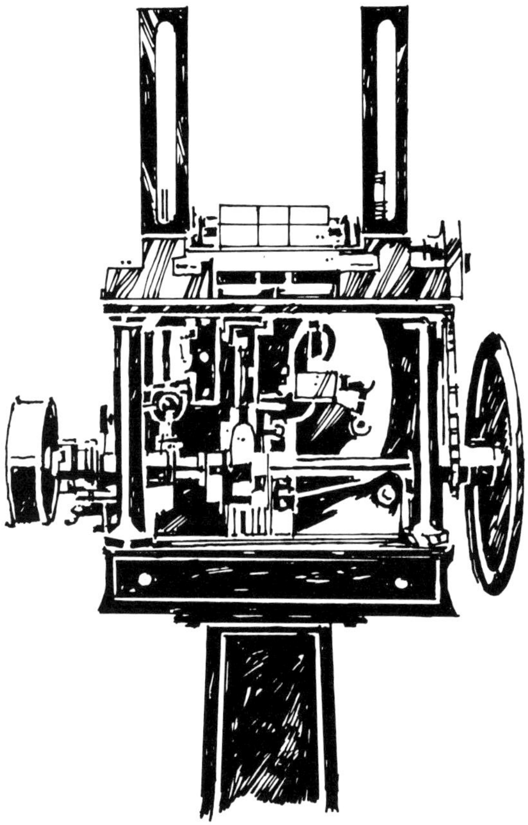

Waterlow standard ticket printing machine as used by the majority of British railways, drawn from catalogue illustration. The price quoted was £150. (Scraperboard by Alan Cassedy)

Waterlow had from the last quarter of the 19th century become one of the main suppliers of ticket printing machines as well as printed tickets. Apart from Edmondsons, their primary rivals were the German firm of Goebel, who operated from Darmstadt and supplied machines to several British printers including Williamson and McCorquodale. Other designs of printing machine featured in American literature include Church & Goddard, Hoe and Rand & Avery; Baranowski is listed as of French origin.

Henry Booth (Hull) Limited, who are now part of the Bemrose group and had previously absorbed both Williamson and GNP, still produce edmondson style card tickets using a Goebel machine with a facility to print both sides of the card in one pass and even remove the half-snip portion. A Goebel machine printing on only one side was presented to Glasgow Transport Museum by GNP, which is now independent again following a management buy-out.

Until 1989 other commercial printers supplying edmondson size tickets (including Bemrose, Bourne Press and J. B. Hacking) generally produced them on conventional machines, printed in multiple on a sheet and trimmed to size afterwards. Hacking also operates Waterlow machines.

With the growth of the railway preservation movement from the mid 50s some unusual sources were found for ticket supply, ranging from local printers to the Eton College Press. One producer who deserves special mention is the late Roger Burdett Wilson of Cheltenham, now widely known for his definitive history of Great Western advertising material. He printed edmondson tickets for several preserved and miniature railways in which he had an interest, including the Great Western Society, Dean Forest Railway and some special events such as annual dinners of railway societies. Much of his collection of ephemera was acquired by Bristol Museum.

Printing by the British railways

The major British railways each had their own printing works. Before the 1923 grouping the Southern's major constituents, for instance, printed at Leake Street, Waterloo (L&SWR), Joiner Street, London Bridge (LB&SCR) and St Thomas' Street, London Bridge (SE&CR). The latter two were combined in the mid-1920s and all were evacuated because of the outbreak of war to Dorking in 1940, where they remained until nationalisation. Railway property in Leake Street was demolished in 1990 in preparation for the Eurotunnel rail terminus.

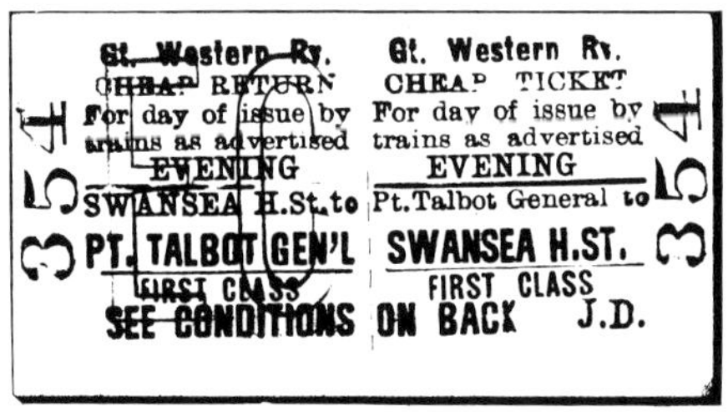

So far as the remaining three of the "big four" are concerned, the London Midland and Scottish Railway latterly printed its tickets at Carlow Street, Euston, the Great Western Railway at Paddington and the London and North Eastern Railway at Stratford, East London. The LMS had a subsidiary works in Glasgow to print tickets for Scottish stations and until the 1930s the LNER maintained a ticket printing shop at York.

After nationalisation (in 1948) British Railways decided to concentrate all its edmondson type card ticket printing at Carlow Street until 1967 when production was moved to Crewe, where much-modified Waterlow printing machines produced each year more than 300 million tickets. Electronic checks added to the basic machines guarded against errors in numbering and stopped the motor when a pre-set quantity had been completed. The bundles of 250 tickets, once screwed up tightly in a clamp for a string band to be fastened, were strung by machine, and Crewe would despatch as many as 1000 parcels of printed tickets each day.

Composing the Type

Originally tickets had been printed from loose type (Monotype) set by hand. When one considers the amount of information on a piece of card 2.7sq in, it is possible to appreciate the small sizes of type often employed. At Paddington the Great Western Railway's compositors set their own initials in the corner of the ticket, so that it is possible to differentiate between those products of H. J. Hicks (H.H.), Arthur Irwin (A.I.), Ernie Bacon (E.B.), Wally Hodge (W.H.), Bill Luck (W.L.), Bill Davis (W.D.) and *(illustrated left)* Jack Dunston (J.D.). Roger Wilson, already mentioned as an independent ticket printer, set his initials, GW style, on most of the tickets he designed, as does the present author.

Frequently-recurring station names were on some lines set in a single slug of type, the LB&SCR employing an unusual mix of upper and lower case letters. When the latter included a descender (e.g. *g, p, q, y*) it was printed as a small capital. This was an additional security against forgery and the brass printing slug could be made narrower to fit more on the ticket. A feature of early GWR tickets is the use of ornate letters **D** and **G** in PADDINGTON.

The former LMS works at Carlow Street, Euston had employed the Linotype system, in which each whole line of type is cast as one slug, and this gradually became the BR standard, greatly easing handling. However, with the move to a central print shop at Crewe serving every station on British Railways, the storage of more than 250,000 type formes, whether Mono or Linotype, would have presented a problem in space and weight.

BR therefore introduced thin printing plates which can be clipped to a spring-loaded block in the metal chase or frame from which the main body type of the ticket is printed. A separate fibre plate carrying the overprint (actually printed *before* the main body text) with, for instance, **CHILD/R/CHILD** can also be clamped to the chase.

Linotype was retained for the station of origin tickets printed for London Transport and for many Scottish Region issues which, being generally to an identical layout, could quickly be made up by the machine operator from racks of slugs carrying station names, ticket titles and (where necessary) fares. Latterly the type for these tickets was set in Monotype taped together. A professional printer would not have considered such a ploy, but on the Waterlow machines it worked successfully!

Each block or forme of type and every plate carried a reference number for easy identification when handling requisitions and in the type library. Before being passed to the printer each was checked for accuracy and particularly any changes specified - such as revised fares or alterations to station names.

For many years tickets sold at ordinary fares (as opposed to cheap day returns, etc) had legally to show the correct fare, and a method of altering plates was devised to prevent the need for wholesale replacement whenever fares increased - all too frequently!

Station names on platform tickets in particular are often printed off-centre, showing where a prefix or suffix (e.g. Barnstaple *Junction*) has been removed from the plate following closure of a branch line and subsequent renaming of the junction station.

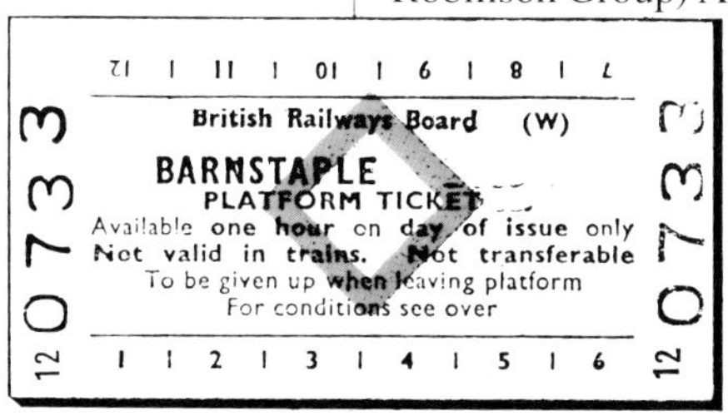

Ticket Boards

Accuracy in cutting card to the 2¼ x 1³⁄₁₆in Edmondson size is essential for smooth feeding into the machine, as is the thickness of the board.

John B. Edmondson recorded that in the mid 19th century the many colours of card for tickets was purchased, ready cut, from Messrs De La Rue, best known as security printers of bank notes and postage stamps and from the Manchester printers Bradshaw, famous for their timetables.

Until nationalisation card generally had been bought in sheets with any conditions wording on the back printed before delivery to the railway. These large sheets were first fed into stripping machines, which delivered the board as 2¼in wide strips into horizontal chutes at the front of the machine. The strips were then placed in a hopper on top of the chopping machine to be automatically fed through rotary cutters and delivered as singles into chutes, at a speed of 90,000 tickets per hour.

As part of the rationalisation in preparation for the move to the central ticket printing shop at Crewe, it was reckoned that it would be more economical for the paste-board to be supplied ready cut to size by the mill. A contract was established with the John Dickinson (later Dickinson Robinson Group) Apsley Mill near Hemel Hempstead, which stood by the side of the main line from London to Crewe. The majority of ticket blanks for BR were supplied with a range of different conditions printed on the reverse; on platform ticket card the red or green rhombus was also pre-printed on the obverse.

A specification issued in 1959 for tickets for the then new Multi- and

Flexiprinter machines gives the dimensions as 2.25 x 1.2in (57.15 x 30.48mm) with a tolerance of +.01in on the width and +.02in on the length; the thickness is quoted as .028in (± .002in). The material is described as *lined ticket middle with lining smooth side outward. Lining paper 60gsm pure* MG *sulphite; middle repulped waste paper, 350gsm. Total substance 470gsm.*

Special Types of Ticket

Railways have always followed an aggressive advertising policy and it is surprising that more use was not made of the backs of tickets for self-advertisements on tickets. There are isolated cases of this form of publicity, in particular the Metropolitan Railway. Several main line companies also used this form of advertising for a time, especially the L&NWR, as did many of the more impecunious independent lines - no doubt to help finance the cost of the tickets. Indeed, one type of Wantage Tramway ticket used more space for advertising than the ticket copy, although this was not of edmondson style.

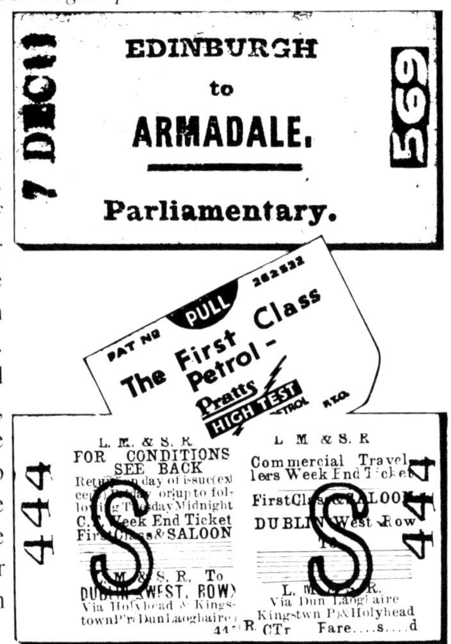

In September 1925 a patent (No 262,522) was enrolled by Messrs A. J. Henderson and E. H. Clifton for making an edmondson style and size ticket from three laminated layers of card scored, folded and glued in a particular way, instead of the traditional pulpboard with facing paper. In punching and trimming the blanks, a small tab was formed enabling the recipient to remove a piece of card about 36 x 22mm which could carry an advertisement. At the same time they patented an alternative design which allowed for an envelope shape to be formed to take an inserted card, similar to an idea patented by Frederic Hall as early as 1868.

In the early 1930s a Mr H. Mones-Cross sold a slightly modified version of the three layer design to the LM&SR, whose tickets carried insert advertisements for such diverse services as opticians, Prudential life assurance and a cancer hospital appeal, and products of the day including Pratt's petrol, Hoover, Typhoo tea, Whisky and Schweppe (sic), Bass and Worthington. The mauve card, used generally for dog tickets, appropriately contained an advertisement for Spratts dog food and Ovals! The LMS also had some self-advertisements. Although the manufacture of this type of blank appears to have continued for only a year or two, tickets survived at some stations well into the nationalisation period.

Many tickets of the last century are described as *Parliamentary* and were issued to travel by those trains operated to comply with Gladstone's *Regulation of Railways Act 1844*. This required companies to operate at least one train a day over all their lines, carrying passengers at fares not exceeding 1d per mile. There were also stipulations regarding speed and type of carriages, but naturally these trains proved to be the very slowest and used the most basic rolling stock permitted by the Act. As normal fares were gradually reduced, the standard third class fare met the requirements of the Act in most cases.

Present day travellers often complain of the plethora of fares between any two points - singles, open returns, day returns, savers, supersavers, etc. In the days of pre-printed tickets there were far more varieties, with special prints for emigrants, police on duty, theatrical parties, walking tours, hop pickers, picnic parties, various categories of military personnel and so on.

British Rail showed considerable initiative to fill empty off-peak seats on the London to Liverpool line by offering fares for *The Executive's Lady*. The plan was that the executive travelled at full first class fare to an early morning meeting in London and his lady used a lightly-loaded midday train to Euston, where they could meet for a meal or theatre visit; they then travelled home together. Unfortunately the offer fell foul of equal opportunities legislation and had to be withdrawn!

Certain suburban journey tickets in London and Glasgow began to appear with brown oxide coated backs, capable of magnetic encoding. The intention of this Automatic Revenue Control (ARC) was for passengers to operate barriers on entering and leaving the platform. London Transport invested in equipment to encode tickets on issue and in many hundreds of gates, operated by inserting a valid ticket, but soon after installation used them only for egress, reintroducing manual control of exits.

BR introduced in 1972 similar barriers in the Gourock and Wemyss Bay lines out of Glasgow, edmondson size tickets being encoded at Crewe during the printing processes. Both entry and exit to platforms were gate-controlled, the innovation here being the use of a stored-journey system. Return tickets were valid for two journeys in *either* direction; weekly seasons were replaced with ten journey cards and a variety of units was available, up to fifty journeys. Each time a ticket was inserted in the barrier a "life" was deducted from the encoding and a digital display showed how many journeys were still valid. When the stored journey experiment ended in October 1979 a BR spokesman commented that "although stored journey principles proved convenient to many travellers, they were troublesome electronically". There was also the fear that the canny Glaswegians had stockpiled the undated tickets before fare increases!

Plans for ARC at four selected Southern Region stations on the Richmond line into Waterloo were published in 1981, but were overtaken by proposals to modernise fare collection all over British Railways. However barriers have been installed as part of the UTS scheme at central area stations on London's Underground, as well as being used on the Glasgow Underground and (for a time) on the Tyneside and Wear Metro.

In the Booking Office

The booking office received its name at the dawn of railways due to the practice of recording passengers' names in a book - and it remained in use until British Rail's new image of the 1960s introduced standardised signs and terms - including the more appropriate (if less romantic) title *Ticket Office*.

Through the years the size of the offices and the equipment provided varied according to the traffic normally on offer. A photograph of Leicester booking office which held a huge stock of different varieties shows a small bench-step provided so the clerks could reach the top rows of the wall-to-wall ticket racking. At the other end of the scale a halt like Pilning (Low Level) near Bristol needed only a small cupboard to hold its complete ticket stock.

Suburban railways and London's Underground in particular called for another variation. On some lines flat fares were charged, but as distances grew and interchange between lines brought through fares, space in the tiny "hole-in-the-wall" offices was at a premium. To alleviate matters scheming or zoning was introduced so that one ticket value would suffice for many destinations.

In an effort to prevent congestion and to speed the flow of passengers, coin operated machines were linked with turnstiles, termed *passimeters*, released as the coins dropped. A development introduced at Kilburn Park in 1921 was to site a manned booking office at the entrance, with a passimeter released at first through an interlocking system with a machine in the office which removed a small round piece from the base of a ticket. Later the gate was manually released, but in either case ticket sales and inspection could be carried out at the same time by a single person.

The passimeter offices remained in use at many Underground stations, latterly without the gate, until the computerisation of fare collection, by which time it was deemed necessary for security reasons to revert to the hole in the wall office, with even the coin operated machines being sited so they could be restocked and the cash removed from behind, away from the public gaze.

In the short period between the end of the war and nationalisation the railway companies began to revive pre-war modernisation plans. The LNER, for instance, picked up the threads of their electrification plans for the London suburban services and at the same time conducted research into the design and modernisation of ticket offices.

Stratford, Maryland and Bowes Park were among the first stations to be given a new look as post-war building materials restrictions allowed, although they were not completed until the LNE had become the Eastern Region of British Railways. From the public's point of view the most visible change was that the tiny keyhole booking office window was replaced by a wide plate glass panel through which the hitherto secret world of the booking clerk became totally visible.

The staff benefitted from the hopper-and-check louvred booking window, the use of modular office furniture and the grouping of functions into three main areas. At Stratford tickets for bookings to London Transport lines, a few Eastern Region stations and the issue of weekly seasons were immediately to hand as the clerk stood at the counter. Those less frequently required, for distant LT stations and most other Eastern Region destinations were held in racks behind him, whilst work to prepare long-term season ticket issues, where notice is required, could be relegated to the rear of the office.

A feature of the new offices was the replacement of the traditional gravity-feed ticket racks by Bellmatic equipment supplied by the Bell Punch Company of Uxbridge. This reverted to Edmondson's original principles of having the next ticket to be issued at the top of the pile and readily visible; there the similarity ended, for the Bellmatic tubes used internal springs to push up the pile of tickets. The cadmium finished tube unit (later replaced by plastic) held 100 tickets and the main problem was that the bundle of tickets had to be numbered with the lowest figure at the top, instead of at the bottom as previously. The difficulty was overcome by having new number print wheels engraved in opposite order and to begin printing at the highest number and working backwards!

Racks of this design were installed at stations throughout the south-east, particularly after the Southern Region introduced cash registers for validating and dating edmondson-size tickets which were numbered only at the top. The last major station to change from gravity-feed to Bellmatic racks was Brighton - although it was not long before even this system was superseded by new technology.

Mechanising Ticket Issue

THE main drawback to the edmondson system was its sheer size, with every station needing a stock of differently-printed tickets for every journey normally taken. This total had to be multiplied by first, second (and formerly third) class issues, adult or child tickets and singles, returns, day returns, saver and other special returns. There was also a risk that a supply of tickets could be stolen on route from the printers to a station, and then used fraudulently.

Machines which printed tickets in the booking office on blank card of edmondson size, and recorded the number and value of each ticket sold, were in use on the continent by the early years of this century. These included the German-designed AEG Regina machines, a number of which were installed by the GWR at Birmingham Snow Hill in 1911, followed by further similar units in the 1920s.

Later machines of similar design were used by the LNER at Newcastle (from June 1931) and Liverpool Street, London (from May 1935). The Newcastle machines were from AEG and used edmondson size ticket blanks, but the Liverpool Street machines differed in being provided by the Westinghouse Ticket Machine Co (a subsidiary of the long-established signalling specialists of Chippenham, Wiltshire) and in issuing tickets produced from rolls of blank card. A smaller machine (similar to the later Flexiprinter) was also tried on the London Underground at Baker Street and Edgware Road from 1938. The Underground also used coin-in-the-slot and other automatic machines to speed the flow of passengers, many of the former issuing edmondson tickets.

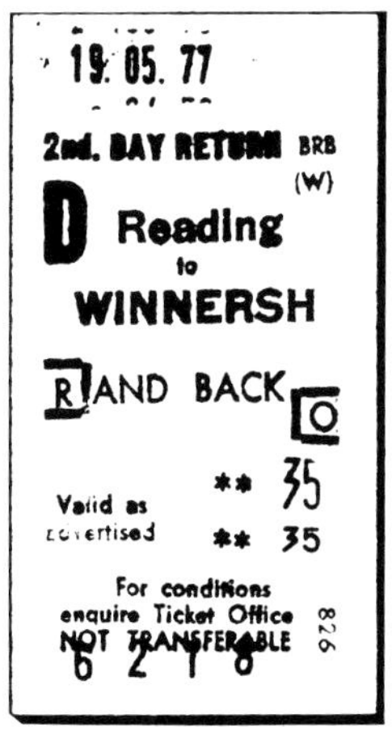

Economies necessitated by the outbreak of the second world war ensured that the edmondson system remained virtually unchallenged on Britain's railways until the mid 1950s when BR began to mechanise. From 1957 many stations received machines produced by Westinghouse, which included the updated Regina, now renamed the Multiprinter, and produced in two versions, capable of issuing either 630 or 1260 different tickets. A cursor to select the ticket required operated across an illuminated screen and the ticket printing was electrical, instead of the mechanical-operation of the 1911 machines.

Where large numbers of a limited variety of tickets were required the Westinghouse/AEG Flexiprinter was installed, offering a printing speed of 0.8 sec or 120 per minute if pre-printing for individual issue later. A print block was taken from a rack and placed in the machine before being twisted to print one or more tickets on blank cards (of edmondson size) held in a carousel. The machine could also print season tickets on slightly bigger card 2.75 x 1.81in (69.85 x 45.97 mm), though not as large as the traditional British season. Widely used on suburban lines around London, the largest Flexiprinter installation in Europe was at Cardiff to cater for the *Valleys* lines bookings.

A third Westinghouse machine, the Handiprinter, produced tickets from a roll of (usually) yellow security-grid paper and was favoured by the London Midland Region and to a lesser extent the Western Region. Many booking offices were provided with Rapidprinter or the generally similar Miniprinter machines, as on London's Underground, for their most popular bookings. By depressing a button on the counter, tickets were disgorged by this machine at a rate of 200 per minute and it is recorded that one window issued 663 tickets in 30 minutes.

At many stations the most popular issues were covered by paper tickets of approximately edmondson size (2½ x 1³⁄₁₆in) dispensed from Bell Punch Ultimatic machines. The tickets were supplied, pre-printed, in zig-zag packs of 1000 with sprocket holes punched in the centre and at the ends. Each machine held five different varieties and depression of the operating lever turned a sprocket wheel, which engaged the punched holes so that a ticket emerged from a slot at the front

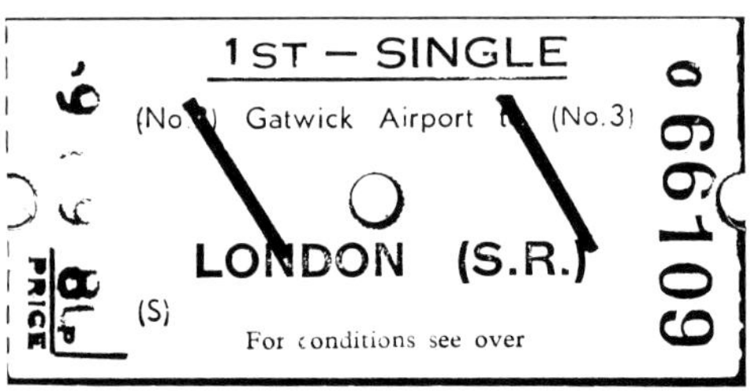

of the machine, ready dated. As fares increased rapidly the last two digits (intended to print the year) were used for the pence element of the fare; for tickets costing more than £1 the pounds were pre-printed on the ticket. At Leeds and Glasgow Queen Street several dozen machines were to be found in use, banked up at each window.

As pay-trains were introduced on branch lines and stopping services where stations were unstaffed bus type ticket machines were issued to guards. The introduction of on-train fare collection brought about the withdrawal of Ultimatic tickets from Leeds, only *InterCity* bookings being handled at the ticket office and the redundant issuing machines were redistributed elsewhere. At Bristol (Temple Meads), for instance, a range of popular travel and even platform tickets were printed in Ultimatic style and issued from former Leeds machines.

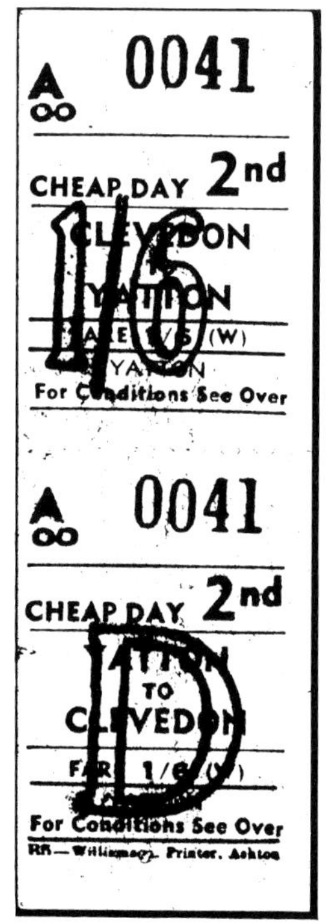

Conductor-guards on some lines such as Yatton to Clevedon, near Bristol, were issued with the familiar rack of pre-printed punch type tickets which had been used since the 1900s on rail motor and auto-trains, but in other places Setright Speed and Almex machines were introduced. Guards on the four-wheeled railbuses used on the Western Region's Kemble to Tetbury and Cirencester branches used special Setrights which printed the name of the starting station or halt in abbreviated form. One of these stopping places was Trouble House Halt, which became well-known because it served only a nearby pub of that name and on the occasion of the final steam train someone had pulled the communication cord to enforce a stop there while the bar was emptied of alcoholic refreshment!

Where more detail was required on a ticket, the Westinghouse/AEG Omniprinter was often used, as in addition to the fares this printed a code letter of ticket type and a two figure code for stations from and to. There was also sufficient space on the fixed printing plate for lists of tickets types and station codes, and the latter could be pre-printed on the back of the ticket roll as well if required. Omniprinters and Almex machines were also provided in the ticket offices at some smaller stations, mounted on a cradle on the counter.

Beginning in 1966, NCR21 cash registers were used on the Southern Region to validate pre-printed edmondson size tickets. They inserted the date, fare and machine code number, while a totaliser mechanism allowed speedy cashing-up at the end of a shift.

Other, much more sophisticated, cash register type machines were used elsewhere, capable of printing originating station, ticket type and fare on large, partially pre-printed tickets and providing a record to be electronically processed. These machines were the Hugin, first installed at Kings Lynn and Sheffield in 1967 and later at a few other stations, and the NCR51, installed extensively from 1971, mainly on the Western, Eastern and Scottish Regions.

A major advantage of this model was that common ticket stock could be used for all issuing stations, as the destination only was pre-printed and the name of the issuing point added by the validation. Also these tickets did not present any

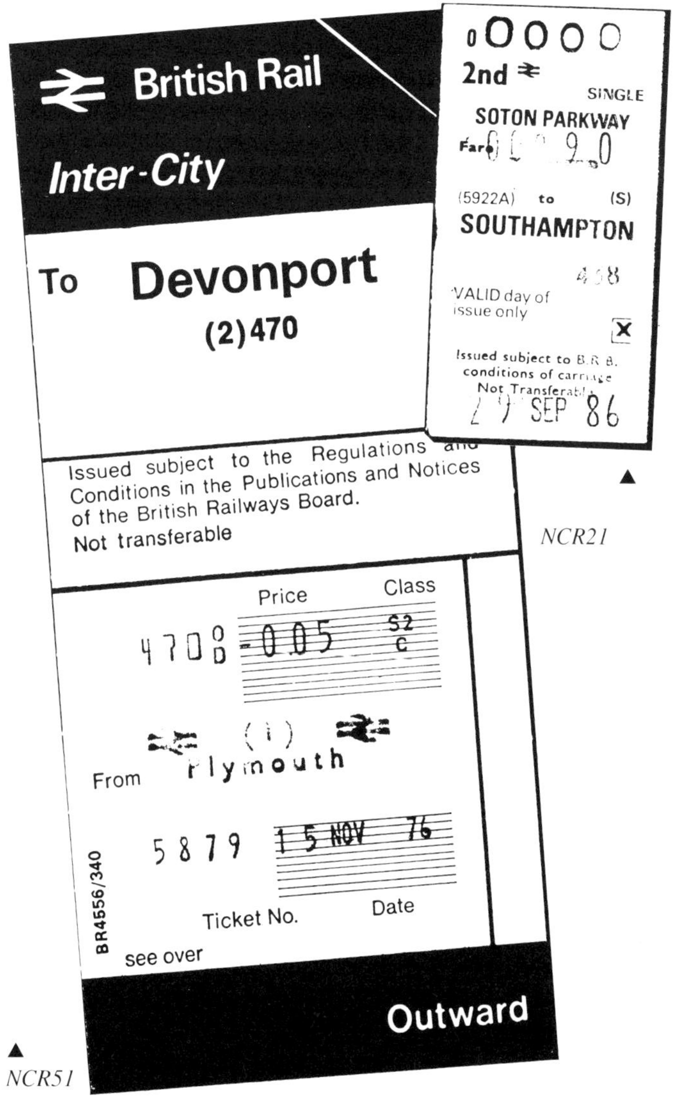

▲ NCR21

▲ NCR51

security risk whilst in transit or in stock, since the cards were worthless until run through the machine on the booking office counter. An anomaly was that all tickets were printed with the heading *InterCity*, even when used for short-hops on suburban services!

Despite these developments, British Rail's Crewe works continued to print as many as 80 million edmondson size card tickets each year. Indeed, the Southern Region NCR21 and NCR24 machines used tickets of this traditional size. Other pilot schemes included the stored-journey system with magnetically coded tickets and passenger control barriers in Glasgow and, in 1979, the Small Stations Ticket Issuing System (SSTIS) using machine-processed credit card size tickets pre-printed with destination stations only. This equipment was developed by Westinghouse-Cubic into INTIS (Intermediate Ticket Issuing System) installed at selected stations throughout the country. *(Tickets illustrated below: left INTIS and right APTIS.)*

In 1984 British Rail announced they planned to make full use of microchip technology by introducing computerised ticket issuing at all staffed stations and on trains, an estimated investment of more than £30 million with most of the equipment being developed in conjunction with and manufactured by Thorn-EMI Ltd at their Wells, Somerset, factory. The new system was to be known as APTIS (All Purpose Ticket Issuing System). As it was refined and made more sophisticated to include additional ticket office functions the interpretation of the acronym was changed to Accountancy and Passenger Ticket Issuing System.

Desk-mounted consoles with a keyboard are provided at all ticket office window positions to issue credit-card sized tickets printed with a security grid background and punched with rounded corners to ease handling through the machine. The full range of fares from the home station can be called up and discounting for railcards handled at the touch of a button.

A hopper in the machine holds a stock of standard ticket blanks; special issues for seasons, car parking, reservations and charter control could be hand fed. Further refinements introduced later, included holding a range of fares from nearby principal stations, automatic credit card wipe and voucher issue, and coding of a magnetic strip on the reverse. This can be used by travelling ticket inspectors for checking with hand-held verifiers and for operating passenger control barriers as used on London Underground.

During the night the machines are linked automatically through railway telephone lines to a central computer in Nottingham. Data regarding the day's sales is retrieved and any fare changes or new facilities downloaded from central control to the station. For short-term or local promotions, data can be entered by the clerk using bar codes and a wipe-pen.

Associated equipment includes PORTIS (Portable TIS), a compact machine designed for on-train issue and carried by conductor-guards on a neck-strap. It offers a fully-printed paper ticket produced on a partially-preprinted roll with security background. The portable machines were sold also to independent lines, including the Romney, Hythe and Dymchurch. The Northern Ireland Railways use the PORTIS machines at both stations and trains, and British Rail has introduced them at ticket barriers and in offices of stations staffed only part-time.

An important feature of the fare collection modernisation plans was the installation of passenger-operated ticket issuing machines in an effort to speed the flow of passengers and reduce staffing. Allocated the overall acronym of POTIS, equipment from several companies from the UK and abroad has been installed at many hundreds of stations, mainly in suburban or heavily-trafficked areas. Some are totally unstaffed, whilst at other locations they supplement ticket offices, particularly where a station is staffed for only part of the day.

A wide range of destinations and types of ticket (including railcard discounts) can be dispensed from the majority of current machines, which are more compact than the bank of three Landis & Gyr Sodeco machines introduced at Waterloo in 1973 for destinations on the Southern Region. They are, too, a far cry from the cast-iron pull-bar platform ticket machines once familiar at station entrances. These machines were once used to dispense low fare tickets on London's Underground and the Great Western Railway used other primitive machines in suburban areas, with tickets printed on extra-thick card (and overprinted A for automatic).

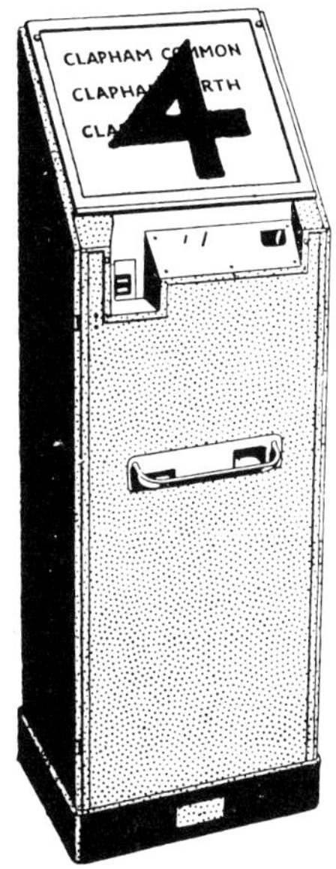

London Transport invested heavily over the years in electrically-powered passenger-operated ticket machines, a long-lasting and familiar design being built by Brecknell, Munro and Rogers Limited of Bristol. The casing, with an angled illuminated screen indicating a list of destination stations became as familiar as the modern LT station architecture, and the machines were speedy and efficient.

Demise of the edmondson system on BR

The completion of the Accountancy and Passenger Ticket Issuing System programme signalled the demise of Mr Edmondson's invention on the main line railways of the United Kingdom and the diminishing production of tickets at Crewe by BR is illustrated in the accompanying chart.

The last full-time employees left the Crewe ticket printing works at Christmas 1987. Mrs Pat Howell, the final operator, expressed her sadness to a reporter from the *Financial Times* and recalled past Christmasses when the whole print shop would have been decked with paper chains, holly and mistletoe. With her to the end was Mr Bob Bradley who, as engineer, had nursed the geriatric printing machines, all of which dated from before railway nationalisation in 1948; cannibalisation meant barely six of the thirty six were operable by then.

The solitary machine which had not been modernised on the move to Crewe and had been used for printing backs and overprinting was put into working order and delivered to the National Railway Museum at York for their collection. They already held the remains of a J. B. Edmondson ticket printing machine formerly displayed at the old York Railway Museum on the pretext of having been used to print souvenir tickets for the 1925 Centenary of Railways celebrations.

During the next few months former employees were occasionally transferred from duties elsewhere in the Paper and Print Department to run small quantities of tickets to top up stocks at isolated stations and travel agencies. As the programme to instal APTIS equipment was nearing completion a few stations remained unchanged, usually due to the rebuilding of premises being in progress.

Thus Emerson Park on the Network SouthEast Romford to Upminster branch became the final BR station to hold a full

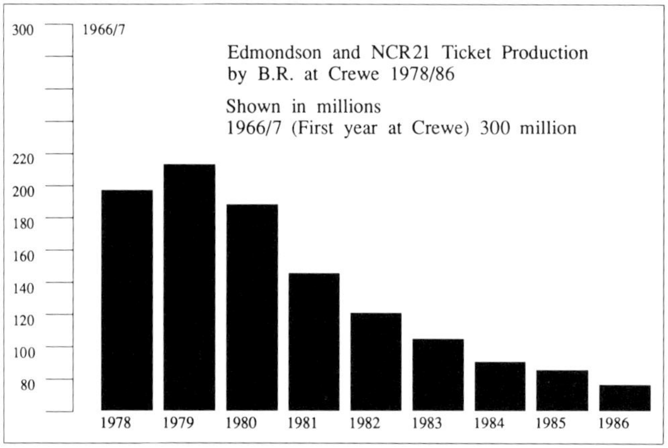

range of edmondson tickets, which were last issued there on 29 June 1989. Eleven months earlier, on 29 July 1988, the last traditional size ticket to be dated and priced in a Southern Region NCR21 cash register had been sold at South Merton station, served by Wimbledon to Sutton trains.

London Transport was concurrently introducing its UTS - Underground Ticketing System - and the final station to issue edmondsons was Covent Garden, UTS being brought into use there on 1 July 1989. LT had pursued a policy of reprinting tickets to carry revised fares when increases were introduced and in 1979 British Rail printed 67 million tickets for them, but two years later this had dropped to 36 million. (BR instructed staff to amend increased fares on ordinary tickets manually as they were issued.) Anticipating the closure of the Crewe print shop, LT had transferred the printing of almost all edmondson tickets to Bemrose of Derby, who had for many years supplied many of their season tickets.

The wooden racks of tickets, once familiar on the back wall of every branch of Thomas Cook Limited and other national travel agencies had disappeared as their shops were re-equipped and the major retail custom came increasingly from foreign holidays and air travel. However the growth of business travel agencies established a need for preprinted tickets to principal destinations and the turnover justified the investment by many agents in their own APTIS machines. At the other end of the scale, many smaller provincial agencies had increased rail bookings where BR de-staffed local stations and several agents operated offices near the stations, some actually owned and operated by former railway staff.

These agents were advised that new supplies of card tickets would cease to be available from 14 October 1988 and that BR planned to issue a pre-printed version of their standard paper manuscript ticket as a replacement. They proved unpopular with agencies, but most, including Halls at Uttoxeter and Wem Travel in Shropshire dutifully withdrew their Edmondson cards on 9 September 1989, as did companies such as British Aerospace which held a stock of tickets for staff travel from Bristol to London and Stevenage.

Emyr Phillips of Pembrey Travel in South Wales managed to persuade the powers-that-be to allow him an extension, but even he had to abandon the time-honoured system and he issued his last card ticket on 4 February 1990. BR Provincial marked the event with a reception at a local hotel and a presentation of the last tickets, suitably framed, to the local mayors and a traveller of long standing. Guests at the Pembrey wake were told it was the end of an era; there would be no more edmondsons for travel on British Rail.

Ironically it was a computer which proved them wrong! Red Funnel Ferries, based at Southampton, offered through bookings to mainland stations from Cowes, Isle of Wight and introduced early in 1990, with BR's agreement, computerised fare collection. Their system was seen as a possible

Issued at Emerson Park on last day of sale - 29 June 1989.

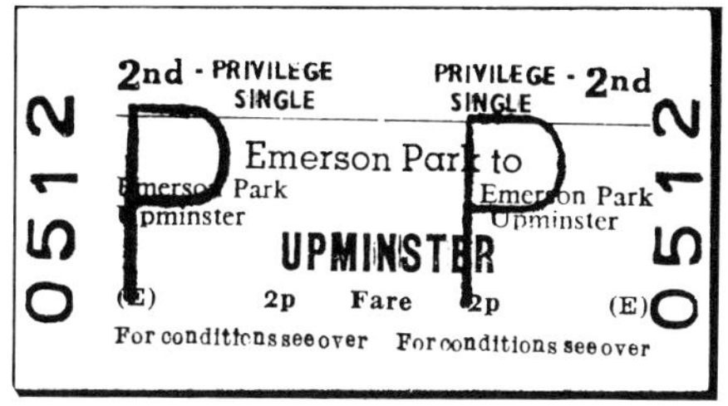

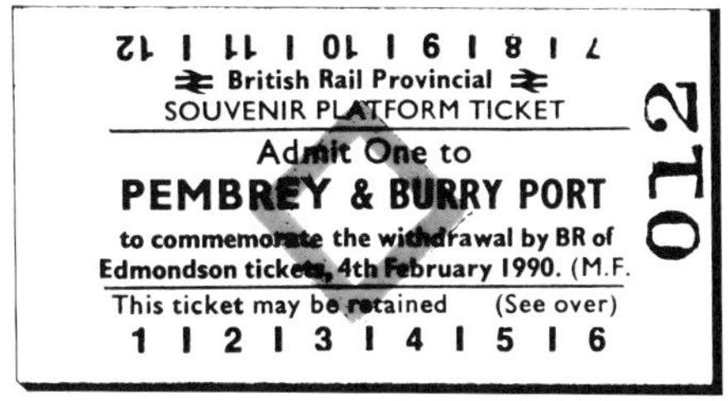

Souvenir platform ticket from Pembrey & Burry Port.

Travel ticket issued by Pembrey Travel on final day before changing to paper tickets.

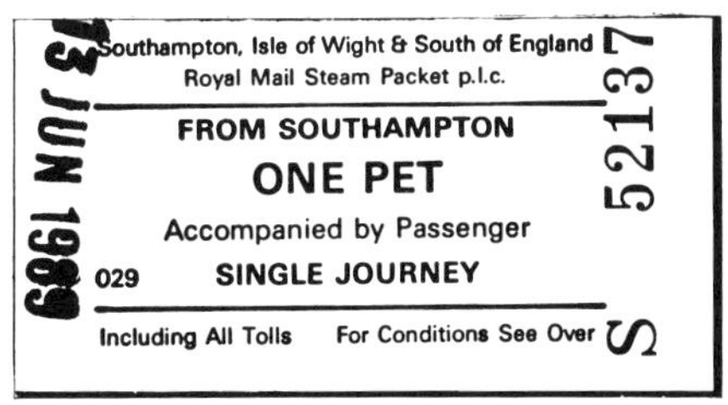

Red Funnel Ferries (note full title printed on ticket) sold this ticket for an unusual class of traffic!

installation for smaller agencies and was to undergo a two year trial at Cowes, where more than 1000 tickets a week were issued, too many to operate the paper ticket method.

Red Funnel's edmondson tickets for through sea and rail travel, printed at Crewe, remained on sale at Cowes, last being used on 24 February 1990. They were held in stock for a little longer in case of problems with the computer, but were not required again and the BR ticket stocks were returned to the Audit Office on 9 April 1990. Red Funnel's own edmondson tickets for sea journeys only had been printed initially by Waterlow, then by Williamson and finally some very attractive tickets - including an issue for **ONE PET** printed (on larger sheets and trimmed afterwards to edmondson size) by Pinecolour Print of Southampton.

Legislation aimed at making BR more commercial (allowing, for instance, for BR Engineering Ltd to tender for outside work) permitted the Crewe factory to print tickets for independent and preserved lines, beginning with the Umfolozi narrow gauge railway at Whipsnade Zoo. Customers ranged from the Severn Valley with nearly 200 different varieties to standard and narrow gauge railways all over the country, as well as model and miniature railways requiring only one type of card, often for a special exhibition. They even produced a wedding invitation in the form of a platform ticket, with BRIDE and GROOM top and bottom in place of hour figures, so the ticket could be clipped appropriately when proffered at the church door by guests.

British Rail did consider retaining a limited ticket printing facility and the independent companies were asked in February 1987 to estimate future requirements. However, the APTIS installation programme was gaining momentum so the policy was not pursued and a further letter called for "last orders". Many lines took the opportunity to order large quantities to last them several seasons, as BR's prices were extremely competitive and they still offered a range of five card colours - white, pink, green, lavender and yellow.

The main competitor at that time for tickets printed traditionally, Henry Booth Ltd of Hull, stocked only white card and printed (on their Goebels machine) stripes or colour blocks as part of the process. They enjoyed the patronage of a large number of independent railways, and secured a large contract for printing some attractive admission tickets for the Motor Show, issued at BR stations in connection with a joint travel/admission promotion.

BR offered the Waterlow ticket printing machines for tender with the stocks of blank tickets early in 1989. Bids were made for the operable equipment by independent railways, commercial printing companies and some individuals, leaving only a few incomplete machines and a variety of spare parts, most of them in very second-hand condition. Examples of the Waterlow machines are displayed at several transport museums, including the West Somerset Railway at Blue Anchor, Winchcombe Railway Museum, Gloucestershire, Amberley Chalk Pits Museum, West Sussex and Haven Street Station on the Isle of Wight.

At the same time the Crewe depot of the Paper and Printing Division of the Director of Supply was closed. Much of the storage space was in a huge factory built by the LNWR for manufacturing and repairing uniforms, where in the first world war military uniforms had been made as part of the civilian war effort. By the late 1980s the building was in a sorry state of repair and BR's decision to hive off most of its stationery supplies to an independent company, Sandhurst, was timely. Internal control was moved from Crewe to Derby.

Some of the staff at Crewe had the foresight to save the remains of the unsold Waterlow machines from the scrapheap and move them to the Crewe Heritage Centre. In 1990 they were reassembled by the former ticket shop engineer, Bob Bradley, to provide an operational machine, with the intention of producing admission tickets for the Centre.

BR found there was a need for edmondson tickets to be printed for special events such as the replacement of Holborn Viaduct station by St Paul's Thameslink, the centenary of the Forth Bridge and the opening of the Ness Bridge at Inverness after the old structure had been swept away. For the traditional enthronement of the Archbishop of Canterbury in April 1991 a special train was provided by Network SouthEast and passengers were issued, appropriately, with traditional railway tickets. For all these happenings BR had to enlist the aid of individuals who had bought their machines from Crewe!

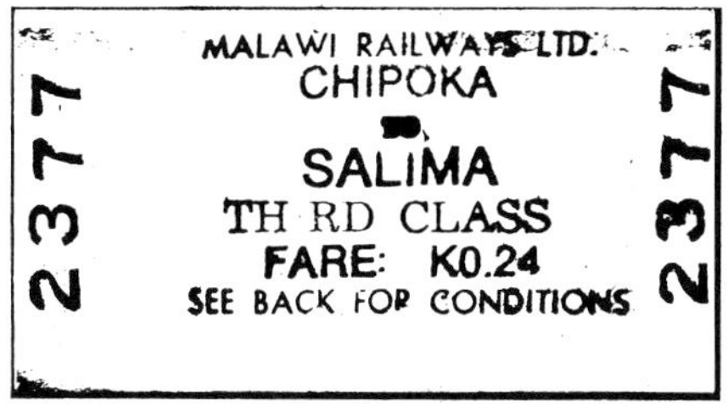

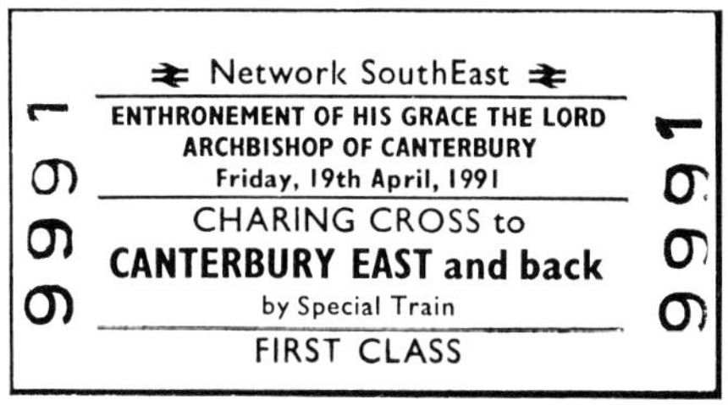

Despite the general withdrawal of all edmondson tickets when APTIS or PORTIS machines were commissioned, a few stations contrived to retain stocks. These were mainly platform tickets - including Manchester Victoria, Thurso and Wick (Britain's most northerly stations) and also, appropriately, Crewe - the former home of BR ticket printing. Feniton on the former L&SWR Waterloo to Exeter line managed to keep for a few extra months a stock of season ticket size day Capitalcards as their PORTIS machine could not issue the special tickets valid for a return journey to London and then a day's travel on public transport in the capital, and passengers objected to exchanging paper tickets at the city terminus.

The Worldwide Scene

Overseas the Edmondson card is far from dead, although mechanisation and computerization has developed rapidly in the past few years. Many African countries retain the traditional system, Malawi having commissioned through the Crown Agents the very last Waterlow machine built - by ERA Ltd of Dunstable and delivered in February 1977.

In Australia, printing of tickets for railways in Queensland was the responsibility of the Government Printer, Mr S. R. Hampson. In superb new offices built at Woolloongabba in the 1980s, beside the very latest printing and reproduction equipment, three geriatric printing machines have manufactured tickets by the million. Two machines had been supplied by J. B. Edmondson in 1912 and a third, by Waterlow, had been acquired from the State of Victoria printers in 1970.

A pioneer of railway preservation in the antipodes, the Emerald Tourist Board's Puffing Billy Railway, was instrumental in saving an 1884 Edmondson machine from Tasmanian Railways and subsequently four further Waterlow machines from the Victorian Government printer.

In Switzerland mechanisation had begun in earnest with the introduction in 1965 of Pautze equipment to the SBB's ticket offices at more important stations and two years later pioneered on the independent Swiss railways by the Bodensee-Toggenburg-Bahn. The Pautze machine is operated by inserting first a curved printing plate and then a

The Swiss ESG system: *1. Central Control Unit; 2. Roll ticket printer; 3. Insert ticket printer; 4. Screen; 5. Keyboard; 6. Customer's price display.*

(Courtesy: SBB)

blank card of edmondson size. Further Swiss mechanisation took place at principal SBB stations in the 70s and 80s, culminating in the ESG system - Elektronische Schaltergerate. With a keyboard and computer screen at each booking office window position the ESG is able to carry out all kinds of transactions, including issue of tickets and seat reservations, validating railcards and multi-journey cards.

Similar equipment is used at major stations of some of Switzerland's smaller railways; other companies have invested heavily in a system known as IGB. Railways, as well as many of the ski-lifts and funiculars use punch and magnetically validated card systems, including the sophisticated SkiData, produced in Austria. Passenger-operated machines are widely used throughout Switzerland, at both staffed and unstaffed stations, and several of those used by British Rail are manufactured in that country.

However, in spotlessly clean and modern offices by the lineside at Bern, printing machines by Goebel of Darmstadt and the Beweg model by Speedomatic AG of Lengnau still (in 1991) produce tickets not only of standard edmondson size, but also the larger size, nominally 70 x 46mm, used for special tickets. In Switzerland edmondson style tickets continued to be issued at travel agencies and are used by other operators of railways, lake steamers, cable cars and even the PTT (Swiss Post Office) post bus services, being printed not only by the SBB but also independent companies in Worb (near Bern) and Aigle.

Erinnerungsbillett

Fahrt ins Blaue

Samstag, 15. Sept. 1990

EBT SMB VHB

00000

The Rhätische Bahn, operating metre gauge trains in the Grisons region of Switzerland, prints its own edmondson tickets at Chur, although its busiest stations have ESG or Pautze machines.

The Emmenthal-Burgdorf-Thun-Bahn also had its own Goebel machine until 1988, when the printer, Herr Moritz Wächli, retired.

In the UK many of the growing number of independent and

Eisenbahn-Modellbautage

1/2 Preis

Spezialbillett

Gültig am 5. Oktober 1986

Dampffahrt

Seeburg ab 10.20

Seeburg

Küssnacht am Rigi

Fr 5.—

00006

preserved lines have pledged themselves to retain the edmondson system as part of their re-creation of the transport scene of days past. However, portable machines such as the Setright and TIM were ideal for on-train issue and the Keighley and Worth Valley bought some of BR's obsolete Hugin validating machines and later NCR cash registers from the Dutch Railways to date and price card tickets of edmondson size.

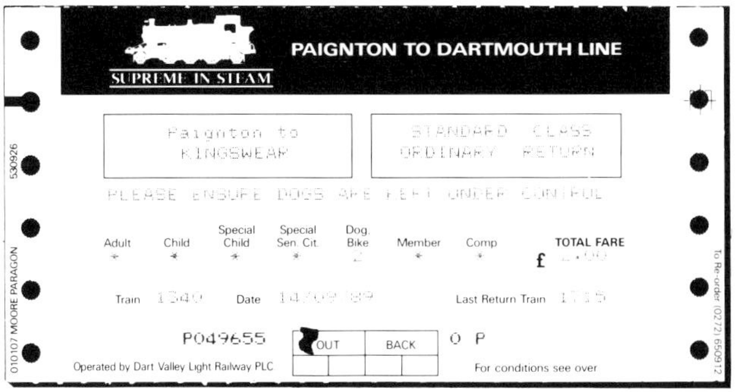

Always an innovator, the Ffestiniog introduced the Titan system, developed by a member of staff. Based on a personal computer, it produces tickets (size: 127 x 76mm) to cover all types of journeys with full cash control and management statistics. The system has also been installed on the Romney, Hythe and Dymchurch, Torbay and Dartmouth and Dart Valley Railways.

Thomas Edmondson Lives!

With the demise of the edmondson ticket in sight, members of the Transport Ticket Society tried to persuade British Rail to name a locomotive after its inventor, but without success. An approach to the Association of Railway Preservation Societies forged links with the South Tynedale Railway Preservation Society who operate the nearest revived line to Milton, the birthplace of Thomas' brainchild. Based at Alston, they had relaid 2ft track along the bed of the former British Rail branch which once connected with the Newcastle and Carlisle line at Hexham.

By good fortune restoration work on their Henschel 0-4-0 tank locomotive No. 6 was almost complete but it had as yet been unnamed. Through the generosity of a TTS member, a pair of attractive brass nameplates was cast with *Thomas Edmondson* and a representation of two clipped tickets. The plates were unveiled by the Earl of Carlisle, the Preservation Society's patron, at a ceremony at Alston station on 23rd August 1987. The railway and the TTS were delighted to welcome two direct descendants of Thomas who had travelled from London and from St Albans to join the celebrations.

In 1990 the operation of the Dart Valley line was taken over from the limited company by the supporters' association under the Buckfast Steam Railway title and the following year this attractive former GWR branch line was leased to the South Devon Railway Trust. Both operators decided to revert from the computerised Titan system to the edmondson card, with tickets produced for them by the West Somerset Railway's Printing Department at Dunster, near Minehead.

Ninteen-ninety saw a double celebration on the pioneer of railway preservation, the Talyllyn Railway - the line's 125th anniversary and the 40th birthday of the Preservation Society. Here, too, there is a commitment to the edmondson card, with a comprehensive range of tickets stocked in the station booking offices from Tywyn to Nant Gwernol.

Long before preservationists took over the Talyllyn was a pioneer of the unstaffed station and on-train ticket issue. A tiny window was cut in the look-out ducket of the line's solitary brake-van, through which passengers from intermediate stations handed their coppers to Edward Thomas, general manager and conductor/guard. He took a ticket (edmondson, of course) from the rack, dated and issued it. The tradition continues, since the TRPS has ensured that each new brake-end coach is provided with a ticket window.

Once ordered from J. B. Edmondson Ltd and Harlands of the Land of Green Ginger, Hull, the main run of tickets are now printed by Henry Booth Ltd of Hull, with small quantities for special events being produced by Preservation Society members. These include the 1982 Royal Train carrying the Prince and Princess of Wales and the 1990 Midsummer Night's Steam when trains were operated round the clock.

Long may the redoubtable Mr Edmondson's wonderful invention continue to serve the public here and elsewhere, as it has done all over the world for more than 150 years.

Rheilffordd Talyllyn

Royal Train

Nov. 25th, 1982

JOHN B. EDMONDSON, MANCHESTER, 19

RAILWAY TICKET DESTROYING MACHINE.

PRICE ON APPLICATION.

To prevent the super abundant accumulation of printed railway tickets, which have already been issued, the Ticket Destroying Machine is indispensable. It cuts them up into small pieces, and in this way makes it impossible for them to be used again.

FOR PAPER TICKETS, TICKET NIPPERS, &c. SEE OVER.

LONDON, GLASGOW, DUBLIN.

The End !

Appendix A

MAKING RAILWAY TICKETS

MESSRS. WATERLOW AND SONS, LIMITED.

We have it on reliable authority that the issue of plain railway tickets last year by the above colossal establishment was over 279 millions, and that printed railway tickets reached the output of nearly 143 millions. . .

From this office more railway tickets are issued than from any other factory in England, and probably more than from any single factory in the world. Not only many of our English lines derive their supplies from this source, but also a large number of Indian, colonial and foreign railways. At the present moment this last branch of the business contributes orders to the enormous extent of 150 millions of tickets a year.

These numbers, though large, yet perhaps scarcely convey an adequate idea of the amount of work they represent. . .every ticket is printed individually, and even cutting up cannot be done in the wholesale fashion adopted in the ordinary card-cutting.

. . .we begin at the beginning. . .the making of the cardboard of which the tickets consist. The material comes. . .in the shape of reams of paper. . .which receives just a thin coating of paste on its way to an operative, who, by combining four pasted sheets, may be said to be the actual manufacturer of the cardboard.

From him it passes to a couple of brawny-armed workmen, who pile it up in stacks and bring down a screw press, by means of which. . .the loose paste and moisture are squeezed out. It is now dried in a heated room for four-and-twenty hours. . .then pressed by hydraulic power equivilent to a weight of 250 tons.

Before printing, parti-coloured tickets — such as striped tickets, or return tickets, where the halves are of different colours — have to receive these tints. This is done by passing them through a machine. . .with bands of flannel kept saturated with the required colour.

Now comes the cutting-up. The sheets are taken one by one, laid on an ordinary bookbinders' cutting machine, and sliced up into strips just the width of a railway ticket, and long enough to make thirteen. The strips are now passed over to boys, each of whom presides at another kind of cutting machine, looking something like a harmonium. . .by means of blades, they are chopped into thirteen "blanks".

In the printing room there are some 22 small but rather complex machines which strike one as having no ordinary share of intelligence. The blanks are put in a kind of square spout, about 500* of them one upon the other, the bottom of them resting on a steel plate.

The machine is put in motion, and a little metal slide gives a dig at the bottom of the pile, just catching the edge of the lowest ticket, thus starting it on a journey of some few inches across a bridge. This bridge leads over to the bottom of another upright square spout.

By the time the ticket has got as far as its own length across the Lilliputian viaduct it would stop, but that another ticket comes behind it and gives it another push, a third comes behind the second, and so on, the firtst ticket afrer two or three prods behind, being pushed into the other spout.

Arrived here, it next receives a jerk upwards from below, the ticket No.2 comes in beneath it, No.3 beneath that, and so on, every little slip of card on its passage across the bridge having received not only whatever printing is required back and front, but a consecutive number, and if a return ticket it has in addition to the printing a consecutive number at each end and a perforation in the middle.

The one feature of this wonderful little machine that strikes one as rather a droll approach to intelligence is that it will pass none but good "blanks" and the moment it reaches an imperfect or malformed piece of card it stops working until somebody comes to put matters right, and then it will start again, digging among the blanks from one spout, and piling up in the other tickets that cannot but set the spectator thinking what a delightfully cheap autumn holiday he might have if he had but the use of one of these clever little affairs, if only for ten minutes!

Here are tickets by tens of thousands for the most outlandish places under the sun, some of them, as for instance in the case of a Ceylon railway, printed in three languages, to suit the convenience of a mixed population. Various Indian languages and dialects, German, French, Arabic, and a host of other tongues are represented, and the colours and devices of the tickets are as various as the tongues.

The printing machine that numbers the tickets, of course, affords the means of checking the numbers; but to insure accuracy, the newly printed slips are transferred to another machine, which, with a rapidity almost too great for the eye, checks the numeration, and enables any inaccuracy to be detected.

The tying-up and packing present nothing very exceptional; but from beginning to end the various processes constitute, if not exactly a work of art . . . at all events a rather complicated and very interesting manufacture.

Reprinted from the *London Evening Globe* by the *London, Provincial and Colonial Press News, April 1888.*

* — *Rather an over-estimate; 250-300 is more realistic*

Bibliography and References

The Early History of the Railway Ticket. A documentary reprint from the catalogue of John B. Edmondson by the Lancashire and Cheshire Antiquarian Society, Manchester.

Illustrated London News, 22 February 1843, p. 117.

Carlisle Journal, 18 July 1851, reprints obituary notice from the *Manchester Guardian*, 12 July 1851.

Household Words, Vol. VI, 1852.

To Whom we are Indebted for the Railway Ticket System? Letter from John B. Edmondson to *English Mechanic and World of Science* No. 697, 2 August 1878. Reprinted in Cross-Fleury's *Time-honoured Lancaster.*

Our Railway Ticket System by Harriet Martineau, *Household Words*, Vol. VI, 1852.

Handling London's Underground Traffic by J. P. Thomas, published by London's Underground, undated c.1920s.

L.M.S. Magazine, Vol. VI, No. 7, July 1929.

Die Fahrausweise by Prof. Lionel Wiener, published by Int. Eisenbahn-Congress, Brussels, published in parts 1930/40s.

The Newcastle and Carlisle Rly by J. S. Maclean, 1948.

The Jackson Collection of Carlisle Library.

Railway Relics and Regalia, Editor P. B. Whitehouse *(Tickets and Passes* by Maurice I. Bray) published by Hamlyn for *Country Life*, 1975.

Railway Antiques by James Mackay, published by Ward Lock, 1978.

Various publications of the Transport Ticket Society, including *The First Fifty Years. (The Early Development of the Railway Ticket)* by D. G. Geldard, 1984.

Tickets by Michael Farr, *Railway World Annual 1985*, published by Ian Allan Ltd, 1984.

Railway Tickets Timetables & Handbills by M. I. Bray, published by Moorland Publishing Ltd.

Railway and Tramway Tickets by Gordon Fairchild and Peter Wootton, published by Ian Allan Ltd.

150 Years of the Edmondson Ticket by Transport Ticket Society members Michael Farr, David Geldard and John Tilly. *Railway World*, April and June 1988.

Collecting Railway Tickets from the British Mainland by Derek Harris, published by Connor & Butler, 1989.

Farewell to the Edmondsons (tickets at travel agencies) by Michael Farr. *Railway Magazine*, March 1990.

"Thomas Edmondson", South Tynedale Railway locomotive No. 6.